Understanding Innovation

Series editors
Christoph Meinel
Larry Leifer

More information about this series at http://www.springer.com/series/8802

Banny Banerjee • Stefano Ceri
Editors

Creating Innovation Leaders

A Global Perspective

Springer

Editors
Banny Banerjee
Stanford University
Stanford, CA, USA

Stefano Ceri
Politecnico di Milano
Milano, Italy

ISSN 2197-5752 ISSN 2197-5760 (electronic)
Understanding Innovation
ISBN 978-3-319-20519-9 ISBN 978-3-319-20520-5 (eBook)
DOI 10.1007/978-3-319-20520-5

Library of Congress Control Number: 2015958362

Springer Cham Heidelberg New York Dordrecht London

Printed on acid-free paper

Springer International Publishing AG Switzerland is part of Springer Science+Business Media (www.springer.com)

Preface

We all witness the tremendous demand for innovation that is occurring in a wide array of contexts, social, technical and organizational. Unprecedented changes in the business environments and social challenges are driving this demand for innovation.

Organizations are nowadays recognizing that innovation does not happen by simply proposing something new, but is dependent on the capability to enact change. In turn, this depends on the existence of what we can term innovation leaders. Innovation Leaders are people who have the right skills and competencies and are in a position to influence points of view, practices, decisions and actions. This is in contrast to the traditional culture of leadership, which has grown around the norms of hierarchical decision making, risk-averse decision making and resource management for efficiency. In fact, innovation is emerging as a new mode of leadership, one that imagines and crafts alternative futures in the context of open and cooperative processes.

Educational institutions around the world have in recent years given birth to a variety of educational programs having the aim of generating the next cadre of innovation leaders. Such programs have originated somewhat independently and from different fields, but share very similar objectives and a common understanding of the central role of multidisciplinarity and creativity in problem solving.

In 2012 and 2013, a selected group of senior decision makers from these educational institutions, together with representatives from public agencies and private organizations, met in Como (Italy) for what has been named the "Creating Innovation Leaders" summit, a "think-tank" of selected individuals for promoting innovation and creating the leaders of such cultural change, and for reasoning about the opportunities that may favor innovation programs (or the obstacles that should be removed) so as to increase the spreading and relevance of such programs.

During 2014, a small group of ten participants to the two summits decided to spend a week together in further discussion targeted to the consolidation of the main results of the 2012 and 1013 summits in the form of a book. The focus of the book is the process of educating (creating) innovation leaders through specialized programs, pursued by leading academic schools, in the hope that such programs will soon become widespread, consolidate, extend, and become viral; the authors felt that their educational experiences were very similar, and they agreed that a greater awareness of these educational experiences is by itself a tangible output that will help the creation of innovation leadership.

Content

The book starts with Part I consisting of six chapters which have been jointly designed by the ten authors, with a rigorous plan of content progression in mind, which moves from (1) innovation to (2) the ecosystems where innovation occurs, to (3) innovation leadership, to (4) the needs of changing education, to (5) a taxonomy of advanced educational experiences, to (6) cases of positive vs. negative innovation leadership in action upon complex problems. We show that a new kind of innovation leadership is much needed, how it can be created, and how it is put in action.

Subsequently, Part II is a collection of invited chapters which describe ten leading academic programs for creating innovation leaders: their objectives, curricular organization, enrollment procedures, impact upon students. Selected programs include four north American institutions (Stanford's d.school, Harvard's Multidisciplinary Engineering Faculty, the Kambar College at Philadelphia University, OCAD's Master of Design on Strategic Foresight & Innovation), five European institutions (Alta Scuola Politecnica of Milano and Torino, the EIT Master Program, Paris' D-School, Brighton's Interdisciplinary Design Program, Aalto's International Design Business Management Program) and the Mission D program at Tongji University. Thus, while the first six chapters provide the theoretical foundations of why and how innovation leaders should be created, Part II gives evidence that such theoretical foundations are already in action in the programs of ten top-level universities.

We next describe the six chapters of Part I in greater detail. In Chap. 1, we question **what is innovation**. At the most fundamental level, innovation is the ability to deliver new outcomes, paradigms, value, and transformations. It involves using many different skills and modalities to creatively frame the problem, generate radically effective options, make strategic decisions, and manage a pathway to effective execution in a way such that no matter how complex or ill-defined a challenge happens to be, a disproportionately effective or valuable set of outcomes is generated.

In Chap. 2, we discuss how the **innovation ecosystem** results from the complex intricacies between pedagogy, practical processes, delivery systems and services. The innovation ecosystem is working well when educational institutions work together with partner stakeholders, such as government agencies, industry, startups, venture capitalists and non-profit organizations, to build new types of platforms and relationships that create impactful outcomes. This chapter also proposes "innovation diagnostic instruments" to measure the degree of penetration of innovation and the presence of key individual roles and collective group skills within the organization, and the overall strength of the ecosystem in terms of relationships and networking.

Chapter 3 dwells into the main topic of this book, **innovation leadership**, defined as the capacity to continually outperform normative techniques and to identify leverage points for delivering scaled transformations which generate new system behavior. The need of innovation leadership is framed in the context of current society evolution which requires breaking disciplinary silos and normative patterns, yielding to an innovation ecosystem with many new actors and their relationships. The resulting innovation methodology brings about the culture of reframing intentionality, developing new perspectives, reframing the goals, creating powerful and transformational strategies and iterating the delivery of impact through an iterative process of ideating, prototyping, testing, and continuous reframing.

Chapter 4 deals with the **changing face of education.** After recognizing that education is challenged by online sources of "commoditized knowledge", the super-fast evolution of technology required by many professions, and the need of continuous education and of deeper interdisciplinary approaches, the chapter dwells upon new participative models of activities in class and develops a qualification framework for innovation leadership that details knowledge, skills and competences for innovation agents, leaders, and gurus. It then reframes the purpose of higher education and discusses new pedagogical models of interaction, motivation, and engagement, from teaching-centric to learning-centric.

Chapter 5 deals with **interdisciplinary innovation programs.** Interdisciplinary programs and courses are being recognized as an important aspect of education, but curricula transformation is occurring in a multifaceted, chaotic way; codification and systematization are needed for a broader dissemination. The chapter offers a taxonomy of concepts (including intra-, inter-, cross-, trans-disciplinarity) and of innovation program characteristics, offering short descriptions of relevant examples and discussing the key issues and emerging questions for the future.

Finally, Chap. 6 takes an **applied view**, by showing examples of complex contexts where a new innovation leadership has led to a positive outcome, and of problems where the lack of a leadership modalities, as discussed in Chap. 3, has instead created damage, concluding with a discussion of emerging challenges that will require a heavy application of the innovation leadership described in this book.

Intended Readership

The book is dedicated to those who feel the need of providing stimuli regarding innovation and innovation leadership, primarily but not exclusively in academia. These include, but are not limited to, deans and provosts of academic institutions, senior managers in private organizations and policy makers in government. Its intended readership also includes anyone who is engaged in promoting innovation within his own organization, and who feels the need to enrich the intellectual and practical toolbox he uses for this demanding and exciting endeavor.

Though most of the editors and coauthors of the material come from higher education institutions, the overall approach is not academic and research-based. In fact, the discussion directly comes from the collective and practical experience gained not only by designing and running educational programs, but also by delivering innovation-focused projects in many types of organizations, and in assisting governmental institutions in defining policy.

Stanford, CA
Milano, Italy
February 28, 2015

Banny Banerjee
Stefano Ceri

Contents

Authors

Banny Banerjee, is the founder and Director of the Stanford ChangeLabs, which has been conducting research to develop the theoretical underpinnings and new models towards an advanced innovation methodology, especially directed towards grand challenges such as climate change and water security. Banny teaches innovation, strategy, and leadership at Stanford University's d.school.

His area of expertise is innovation methodology for large-scale transformations in the context of complex multi-dimensional challenges.

Stanford University, Stanford, CA

Paola Bertola, Phd, is Professor at the School of Design of Politecnico di Milano, where she teaches in the Fashion Design programmes (BA & MSc). She is Co-Founder of "Fashion in Process" research collective [www.fashioninprocess.com]. She is an expert of design processes management and design-driven innovation in "culture intensive" industries, with a specific focus on fashion companies. In 2011 she was awarded the Compasso d'Oro Price for leading the research program Design Research Maps.

Politecnico di Milano, Milano, Italy

John Body, founded ThinkPlace in Australia in 2005, after a number of senior roles in the Australian Taxation Office where he pioneered the application of design to Australia's taxation system. ThinkPlace now has studios in Australia, New Zealand, Singapore and Kenya addressing the challenges faced by governments, non-government organisations and the development sector. John works with very senior leaders to bring innovative solutions to challenges of interest to societies. John holds a Master's in Chaos and Complexity Theory.

Thinkplace, Kingston, Australia

Stefano Ceri, PhD, is Professor of Database Systems at the Dipartimento di Elettronica, Informazione e Bioingegneria (DEIB) of Politecnico di Milano. He has been Director of Alta Scuola Politecnica (ASP) of Politecnico di Milano and Politecnico di Torino (2010–2013). His research work is generally concerned with extending database technology. He is the recipient of the ACM-SIGMOD "Edward

T. Codd Innovation Award" (New York, June 26, 2013), an ACM Fellow and member of the Academia Europaea.
Politecnico di Milano, Milano, Italy

Fawwaz Habbal, is the Executive Dean for Education and Research at Harvard School of Engineering and Applied Sciences. His research is in nano-photonics and innovative pedagogy. Prior to Harvard, he held the position as Corporate Vice President, responsible for research and product design at Polaroid Corporation, where he served as a Senior Research and Engineering Fellow as well. After leaving this position he initiated three start-ups related to imaging semiconductor electronics.
Harvard University, Cambridge, MA, USA

Nabil Harfoush, is Assistant Professor in the Strategic Foresight & Innovation graduate program and Senior Fellow of Strategic Innovations Lab (s-Lab) at OCAD University, Toronto. He is also a Fellow of the College of Design, Engineering & Commerce at Philadelphia University and Adjunct Professor in the Strategic Design MBA program. Nabil previously served as Chief Information Officer for HelpCaster Technologies, MedcomSoft, and Nightingale Informatix. He has consulted for enterprises, national governments, and international organizations, including the World Bank, WHO, UNESCO, and IDRC.
OCAD University, Toronto, ON, Canada

Véronique Hillen, is the dean of Paris Est d.school since 2012. She has developed many academic programs in design thinking, including ME310 Design Innovation in collaboration with Stanford and its international network SUGAR. She has been the academic director of the Industrial Engineering department at ENPC (2007/2012), a leading engineering French school. She is a visiting professor at the Center for Design Research at Stanford. She has also been a professor of corporate strategy at Ecole Ponts ParisTech since 2007. She has published articles and books in corporate strategy and design thinking. Her professional interests include pedagogical action research for capstone projects with design thinking and how to rejuvenate mature businesses with a culture transformation thanks to design thinking.
Paris Est d.school, Champs-sur-Marne, France

Chiara Leonardi, a Research Fellow at the Department of *Electronics, Information and Bioengineering* of Politecnico di Milano, graduated in "Product Service System Design" with a Master of Science. Her research interests focus on products and complex services and systems to be placed in competitive business environments.
Politecnico di Milano, Milano, Italy

Randy Swearer, is Provost and Dean of the Faculty at Philadelphia University. His primary focus is on innovation in higher education during periods of sector disruption. Dr. Swearer has spoken at venues around the world on the topics of transformative change in higher education, new university models, and innovation leadership in academia.
Philadelphia University, Philadelphia, PA, USA

Federica Vacca, PhD, is Post-Doc Researcher at the Design Dept. of Politecnico di Milano and Lecturer in the School of Design of Politecnico di Milano, where she teaches in the Fashion Design Programmes (BA & MSc). She is Co-Founder of the "Fashion in Process" research collective [www.fashioninprocess.com]. Her research interests concern handicraft-driven creation processes for the enhancement of local culture knowledge and design-driven innovation in "culture-intensive" industries.

Politecnico di Milano, Milano, Italy

Part I
Theoretical Foundations

Chapter 1
Why Innovate?

Banny Banerjee

1.1 Introduction

Across contexts, we see a profound need for identifying and framing problems in new ways, and having framed them, finding radically effective ways of solving them. We face challenges that are complex, urgent and critical. The larger challenges facing our civilization, such as overpopulation, climate change, the energy crisis, food security, water security, loss of biodiversity, and massive urbanization, are *wicked and super-wicked problems* (Rittel and Webber 1973; Levin et al. 2012). These massive challenges are the backdrop against which the institutions and organizations of the world, irrespective of their sector, find themselves having to carry out their functions. In an increasingly connected and interlaced world, we find ourselves enmeshed in "systems of systems" (Deleuze and Guattari 2004). We are entering an era marked by great turbulence and change (Thackara 2005). Organizations across sectors face the dual challenge of dealing with rapid change as well as the pressure to participate in solving the larger societal challenges facing us.

The clock tick of change is getting faster and faster, and as a result, organizations find themselves looking for new ways of maintaining a competitive edge, establishing new markets, and conducting business in a rapidly changing ecosystem. But most organizations have been shaped by a thinking that belongs to the previous century, and consequently struggle to transform themselves into agile, proactive establishments, with value systems more appropriate for the future. This is true for any type of organization, whether it falls in the private sector or the public sector, whether an educational institution or a philanthropic foundation.

B. Banerjee (✉)
Stanford ChangeLabs, Cordura Hall, Rm. 104, 210 Panama St.Stanford, Stanford, CA, 94305-4115 USA
e-mail: banny@stanford.edu

B. Banerjee, S. Ceri (eds.), *Creating Innovation Leaders*,
Understanding Innovation, DOI 10.1007/978-3-319-20520-5_1

This crisis is playing out in the most discernible manner in the business sector. Organizations find themselves calcified into rigid ways of operating, making it easy for new incumbents to challenge their established turf due to their superior agility and their ability to structure their organizations around principles more aligned to the future. This presents a risk to established organizations and an opportunity for startups. As relatively recent incumbents like Google, Facebook and Nest have demonstrated, large organizations such as Nokia, Xerox, HP, Microsoft, and Honeywell, enjoying established markets for decades, can no longer afford the traditional level of complacency. In order to maintain their market dominance, they not only have watch out for other large organizations' fiercely contested turf, but they also have to find ways to defend themselves against smaller, faster and more agile companies that have an innovation edge. This phenomenon implies that an organization without the capacity to innovate is extremely vulnerable to incursions from ones that are innovative and can move fast.

Educational institutions have the dual responsibility not only of creating the right conditions that would feed an entire pipeline of much-needed innovation leaders, but also transforming their own institutions to be more innovative in their structure and organizational culture (Frayling 1991). The two aspects are intricately tied. An educational institution cannot hope to create innovation leaders unless it is willing to embed innovation and all that it implies deep into its own institutional value systems.

Organizational transformation is difficult. It involves changing behaviors, cultures, and mental models and reframing deep-rooted theories of success. It can only take place if there exist agents of change with sufficient influence to drive transformation. Organizational or institutional transformation, and the ability to deliver new outcomes in unfamiliar contexts represent a new kind of leadership, which we term "Innovation Leadership". There is a critical need for this type of leadership, and this is the raison d'être of this book.

Innovation is an overused term that is used in a myriad of contexts. It is used synonymously with advanced technology, creativity, novelty, clever solutions, inventions irrespective of their value, and simply something that is new. If we are to create a framework around innovation leadership, then it is first important to define the nature of innovation that a leader is expected to exemplify and amplify. Below we answer basic questions about innovation, its relevance, and its use for addressing new challenges.

1.1.1 What Is Innovation?

Innovation, at the most fundamental level, is the ability to outperform normative approaches by a significant margin and deliver new value, outcomes, paradigms, and transformations (Banerjee 2009). Real innovation requires great creativity but creativity does not guarantee innovation. A crime can be committed with great creativity, but does not produce positive value. Creativity is the ability to come up with ideas or artifacts that are new, surprising and valuable (Boden 2004).

Creativity is a crucially important ingredient for innovation, but innovation brings about a transformation that has value. It involves using many different skills and modalities to creatively frame the problem (Cross 2006; Dorst and Cross 2001), generate radically effective options, make strategic decisions, and manage execution such that despite the complexity of the challenges, a disproportionately effective or valuable set of outcomes is generated. Innovation is also linked to the act of design. A good definition of design is "the act of turning an existing situation into a desired one". (Simon 1968). Innovation is a larger concept that requires acts of design and innovators assuming a "*designerly*" way of approaching challenges.

1.1.2 Why Is Innovation Leadership an Important Organizational Directive?

Any organization, irrespective of sector, whether or not it is aware of it, implicitly has to manage its innovation portfolio. Educational institutions and policy makers need to drive the growth of this capacity since the ability to find innovative pathways through rapidly shifting contexts is emerging as a critical function. And imparting this capacity to the organization can only take place if there exists a set of innovation leaders who know how to go about imparting a culture of innovation across the organization's many functions.

The following is a list of reasons why it is important for an organization to adopt innovation as a core value and go about creating an ecosystem for innovation.

1. Innovation creates the capacity for an organization to bridge the gap between new types of problems and current organizational capacity.
2. The innovation capacity of the workforce, typically an under-leveraged resource, is representative of untapped potential for value creation and ecosystem leadership.
3. There is a need to rethink existing systems and restructure them to meet uncertain and changing future scenarios.
4. Our current approaches encourage solving piecemeal problems but miss the opportunity to generate systemic solutions. There is a need for systems thinking at all levels.
5. Innovation is a way of reinventing one's own organization and "future-proofing" against the threat of a nimble and innovative incumbent.
6. There is universal pressure to achieve more with less.
7. Most organizations compete in established, competitive markets rather than create "Blue-Ocean Strategies" where they can establish market leadership.
8. Organizations tend to create "push" strategies rather than "pull" strategies, built around a deep understanding of the needs of its stakeholders.
9. *Creating* new value rather than simply *capturing* value is a far stronger strategy.
10. There is a great need for defining gainful ways of being a part of solving the larger problems that threaten the stability of our entire civilization.

1.1.3 When Is Innovation Necessary?

At a high level, challenges can be broken down into two categories: ones that are "Familiar" and those that are "Unfamiliar". Known challenges are ones that are well characterized and their nuances are understood and agreed upon. An unfamiliar challenge is one we might not even know exists, or have a complete lack of understanding about its real nature. Figure 1.1 depicts a 2x2 matrix with these two categories of challenges framed against two categories of solutions: known and unknown.

(A) The lower right quadrant represents "solutions looking for problems". For example a sensing technology that allows you to measure micron-level vibration contains theoretical value, but it may not be immediately apparent where it is best applied.

(B) The lower left-hand quadrant is where known solutions exist for known challenges. For example, it might be fairly well understood how to do knee replacement surgery. This type of challenge is addressed through an "assessment and selection" model, where an existing set of solutions is evaluated, and the solution with the best evidence of success for the given problem is the one that is selected.

(C) The top left quadrant is where there is a fair degree of familiarity with the problem, but we do not have a set of known solutions for them. For example, we might understand chronic diseases that are caused by excessive sedentary behavior, but we might not have a solution to change the behavior of people.

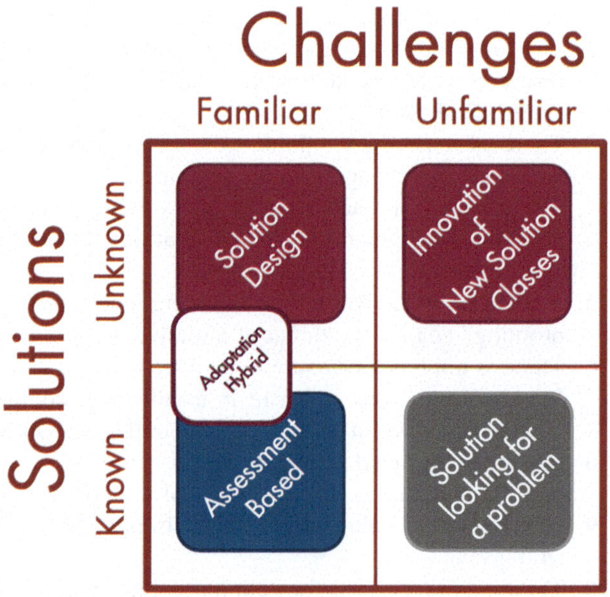

Fig. 1.1 Modified Ansoff Matrix that depicts four different types of innovation classes

(D) The top right quadrant is where challenges are not well understood and there are no ready solutions. Most pervasive systemic challenges fall into this category. Rapid decline in biodiversity is an example of a challenge in this category.

In the bottom left quadrant, since the problem is understood and there exist established solutions, innovation is necessary only if there is a desire to increase performance. In this quadrant, incremental innovation through standard methods is typically found sufficient. In the top left quadrant, sometimes it is possible to innovate through "adaptation" or "hybrid" solutions that involve either taking a solution from a known regime and applying it to an area where there are no solutions, or combining solutions that come from other known challenges. But very often, this quadrant does not lend itself to adaptation and there is a need to innovate new solutions tailored for the challenge type. Occasionally existing solutions can be combined or adapted to a new problem. Even though avoiding the generation of a new solution tends to yield suboptimal solutions, decision makers feel more comfortable selecting from existing solutions rather than a new and optimal solution, thus representing a heavy bias for existing solutions.

The top right quadrant is where some of our most pressing problems lie and arguably where a lot of the other challenges lie that are perceived as being well understood. In this domain, there is no option but to innovate! Not only is there a need for a higher degree of innovation, given the implied complexity (Hayek 1964), since the problem is not well understood, there is a great need in for framing the problem innovatively. Disproportionately effective solutions and new classes of solutions to complex challenges can only follow artful framing of the problem, which in turn relies heavily on deep insights and understanding of the challenge.

The lack of understanding of the taxonomy of challenge types causes a blind spot leading to the wrong kind of solution to be put forth. The logic of the lower left-hand corner is heavily ingrained in our decision-making culture. We condition ourselves to only be able to decide on solutions that have an evidence base that can give us comfort that the solution works. This limits us to only the set of existing solutions and prevents us from exercising innovation capacity in creating new ones.

This phenomenon of seeking evidence before making a decision is such an organizational habit of mind, that decision makers routinely make one or more of the following three profound fallacies when the challenges are not in the bottom left quadrant: (a) the assumption that a solution that works in a known regime can be applied to another regime; (b) the assumption that we understand the nature and the boundary conditions of the problem, and (c) looking for evidence or proven efficacy of solutions as a condition to a decision in a challenge that is not well understood is a valid way of reducing risk. This bias of reducing risk by limiting it to known and tested solutions is a direct impediment to real innovation.

1.2 Framing Innovation

Now let us look at scenarios regarding whether there exists a process to achieve outcomes. The framework in Fig. 1.2, credited to Eddie Obeng, is a truth table that looks at the relationships between the desired outcomes and the processes to achieve them. The four quadrants of this framework are the following:

(A) If the desired outcomes are known, and the solutions to achieve them are codified into a set of steps, then the approach to problem solving is called "Paint by Numbers". The term refers to children's books where you have to join numbered dots for a picture of a clown or lion to emerge. In this case, all you need to know is what the next step is, and the people implementing the solution need not have a meta-picture of problem – they just need to know how to follow an established procedure.

(B) The top quadrant is labeled the "Quest" because in this category the problem is known but the solution is unknown. The problem with this situation is that without a process you could be looking interminably for a solution to emerge, and until you have a solution, you do not have a solution!

(C) The left quadrant, labeled the "Fog", is where one is aware of having a challenge, but it is not clear what the right outcome should be, and there is no process to get you there. If one were to take even problems that are perceived as being well understood and we broaden the boundary conditions or the length of time over which they were considered, they have a tendency of becoming foggy problems.

(D) The fourth quadrant is called the "Movie" project, where there is a very robust process that looks at an ill-defined problem. The process guides people into understanding identifying the appropriate outcomes. It is to be noted that the definition of the outcome is expected to *emerge* as a result of a robust process, and hence the process involves resisting locking at the problem a priori. The reason it is called a "Movie" project is that the movie industry makes highly unique one-off instances that are extremely complex, but the movie-making industry has created a highly sophisticated process that allows it to rapidly take

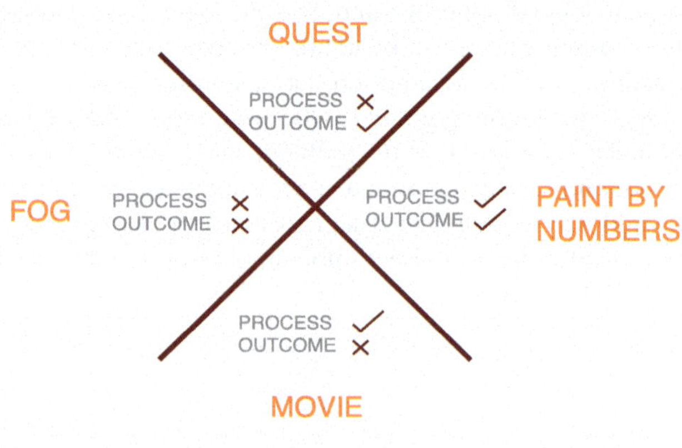

Fig. 1.2 Framework by Eddie Obeng laying out four different scenarios regarding whether the desired outcomes are known and whether there is a process to get to them

very large teams and come up with a highly tangible and multifaceted product irrespective of genre. This framework raises the critical aspect of why you need a process when the challenge type is foggy in nature. The C-K Theory (Hatchuel and Weil 2003) also depicts design as a similar conversion from a "concept space" to a "knowledge space". Since so many of our most pressing problems essentially fall in the category of the "fog", it of great importance that we employ powerful processes that guide leaders and innovators.

The two frameworks combined have some important implications:

1. It is crucial to understand the type of challenge one is facing and not use the wrong approach to address it. If the challenge is in the "unfamiliar" category, then innovation is an imperative since existing solutions are likely to provide suboptimal results
2. If the challenge is "foggy" in nature then we need to rely on an robust innovation process that is well suited for the class of challenge
3. The decision and evaluation criteria need to belong in the regime in which the approach belongs. It is a fatal fallacy to try to base the entire decision of an innovation-based approach, which is "future-based", by seeking evidence from nonexistent or an inappropriate set of solutions.

1.2.1 Innovation Is a Team Sport

One of the core emerging principles of innovation is that a team of people co-creating using innovation methodologies will far outperform the lone genius, especially when it is a complex challenge. Complex challenges are multi-dimensional and tend to defy the confines of a single discipline or an individual point of view. There are different categories of collaboration that are important to make a distinction between.

1. **Inter-disciplinary:** Inter-disciplinary is where the thinking and the tools from one discipline can be leveraged for another discipline or the work is at the intersection of two fields. An example is bioengineering, which looks at the mechanics of biological processes.
2. **Multi-disciplinary:** Multi-disciplinary is where members from different fields come together to solve the problem, but each representative of a discipline operates strictly within his or her disciplinary domain. An example is when a structural engineer, architect, and a landscape designer work together to design a building with the typical division of thinking, where the structural designer dutifully designs the structure without trying to influence the architecture, and the landscape designer exercises control over the landscape in keeping with the larger vision of the architectural statement.
3. **Trans-disciplinary:** Trans-disciplinary is when members of different disciplines come together to "surround" a challenge and use their different

disciplinary perspectives to "co-create" solutions in a manner that requires them to transcend the confines of their disciplines. When trans-disciplinary co-creation is carried out well, it is difficult to discern which discipline might have been responsible for which component. This is a particularly important modality in addressing complex, multi-dimensional challenges. The more advanced innovation methodologies focus on making sure that trans-disciplinary co-creation is ensured.

4. **Trans-agency:** Trans-agency is very similar to trans-disciplinary, except that the boundaries being transcended are not just disciplinary boundaries, but organizational ones. An example might be civic communities, city government officials, small businesses, bankers, social scientists, policy makers, and technology companies coming together to deal simultaneously with the issues of crime and of employment for new immigrants.

There are some other issues that recur when addressing complex challenges:

1. **Multi-epistemological:** Epistemology refers to a theory of knowledge. It refers to the structure by which a field acquires, views, processes, and validates knowledge. The epistemological structure of each discipline tends to be distinct from the others. Even when there is a genuinely trans-disciplinary co-creation in process, there is still a presence of the different epistemologies from the different disciplines that need to be managed and negotiated.
2. **Multi-motivational:** Different disciplines and agencies have different motivations. Rather than trying to conflate every agency's motivation to a single aligned motivation, it is far more realistic to acknowledge that there exist different motivations and the initiative needs to map to, modify, or leverage the multi-motivational space that emerges through a process that brings multiple disciplines or agencies together.
3. **Multi-objective:** Complex challenges are systems of systems where it is not effective to just have one single objective or goal, but instead to create initiatives that result in transformations on a number of different fronts simultaneously.
4. **Multi-cultural:** Often both within the team and in the challenge area, there are multiple cultures and cultural boundaries that need to be understood and navigated. This point is extremely relevant when working with grand challenges that are playing across different cultural contents.
5. **Multi-temporal:** The time frames of different parts of the challenge or different objectives can be different, and it requires thinking simultaneously in different time frames. This is unlike trying to release a product in time for a Christmas sale. For example a strategy for employment diversification for poor communities might lie in a technology-enabled bridge connecting them to established markets. The platform might include an "entrepreneurship ladder", incorporating education and entrepreneurship components, allowing early entrepreneurs to provide employment for the new entrants into the system.

1.2.2 Problem Identification, Not Just Problem Solving

The generalized process of innovation involves navigating two primary spaces: the problem space and the solution space. Since innovation is most necessary in foggy challenges or where existing solutions do not exist or the problem is not well understood, it not just an act of "problem solving". The more important part it to understand the challenge, and identify *what the problem is*. Traditional processes are sequential in nature and tend to freeze the problem domain in a deterministic fashion, casting it into specifications that are then implemented on. Innovation processes are highly iterative. The problem can be reframed in terms of new information that emerges as a result of the innovation process, as shown in Fig. 1.3.

The extremely important frame shift in an iterative process compared to a sequential process is that the understanding of the problem space is achieved through the act of trying to solve it. Iterative oscillations between the problem domain and the solution domain (Dorst and Cross 2001), with testing of rapidly generated prototypes. This approach results in the understanding of both the problem domain and the solution domain advancing in lockstep with higher and higher levels of fidelity.

The most fundamental way of looking at an innovation process is as an iterative loop circling around a central question, as shown in Fig. 1.4.

An example of a challenge might be to reduce the number of babies that die due to complications arising out of premature births and low birth. The key questions to frame could be "How might we save babies dying because premature birth?" "How might we prevent premature birth?" or "How do we design a cheaper incubator?" These are three entirely different questions with completely different classes of solutions arising due to the difference in their framing. In any case, the innovation process then would involve three main activities: (a) generating deep insights, (b) Using the insights to reframe the problem, and (c) conceptualize potential solutions and prototyping them for testing, which in turn would generate fresh insights.

In innovation processes, along with the iteration between the problem domain and solution domain, the perspective also oscillates between a high level of

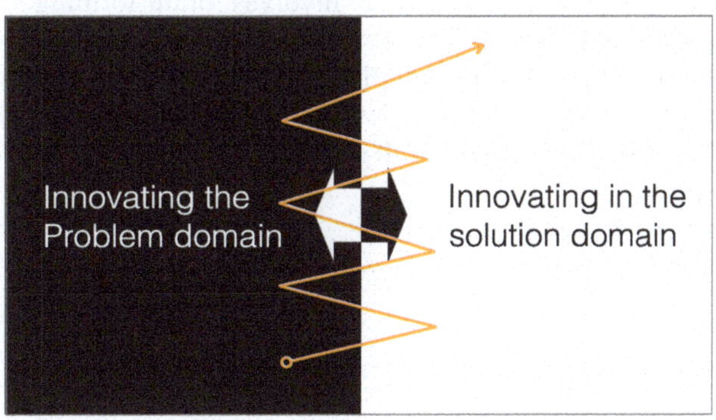

Fig. 1.3 A generalized iterative innovation process depicting moving back and forth between the problem domain and the solution domain with increasing levels of fidelity

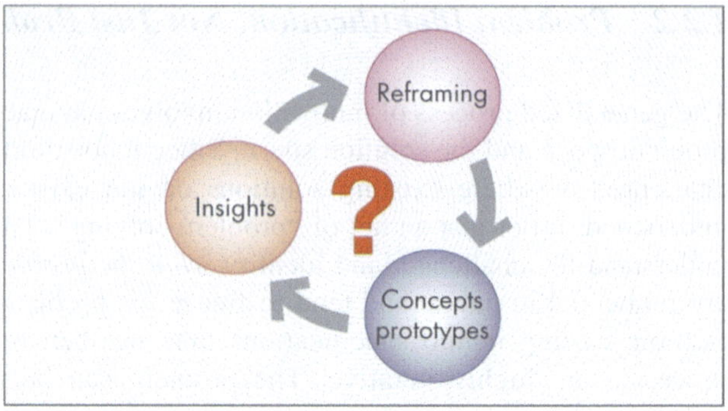

Fig. 1.4 The innovation process as an iterative loop between generating insights, reframing the challenge in terms of the insights and generating concepts and prototypes that can be tested in order to generate more insights. At the center is the fundamental question that defines the challenge

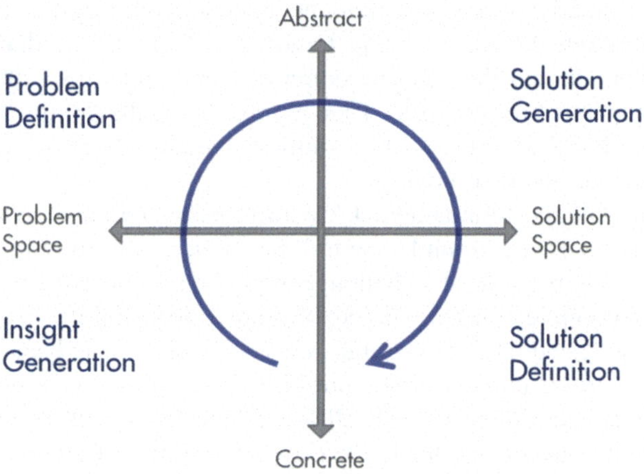

Fig. 1.5 The innovation process as iterative process that goes between abstract and tangible views, and simultaneously between the problem domain and the solution domain

abstraction to a very tangible and concrete level as depicted in Fig. 1.5. The innovation journey begins with insight generation about the real and tangible aspects of the problem, digging deeper into root causes. The synthesis of this information produces a more abstract level view of the challenge. This high level view then allows for a directed but generative phase of Solution Generation. The Solution Generation phase involves brainstorming and concept generation of potential solutions that are then winnowed down into a smaller set of solutions through the act of prototyping and testing. The act of testing gives us new insights about needs, failure modes and system dynamics, and the cycle begins again, but this time, armed with a much more nuanced understanding of the problem space.

The Power of Innovation Process
Reframing infant mortality problem in India
Out of 130 million babies born each year, 20 million are born premature with low birth weight and are prone to hypothermia. In developing countries, four million die within their first month of life. Incubators and radiant warmers are widely used to prevent and manage hypothermia. At the Stanford d.school, in an innovation class, students were asked to think of affordable solutions: "Can you design an incubator at a hundredth of the $20,000 that modern incubators cost?" A four-student team flew to India to visit rural hospitals to find that they actually already possessed incubators, but to their surprise, most of them were empty. Their ethnography research found that families from remote villages had no means of bringing their babies to these regional hospitals.

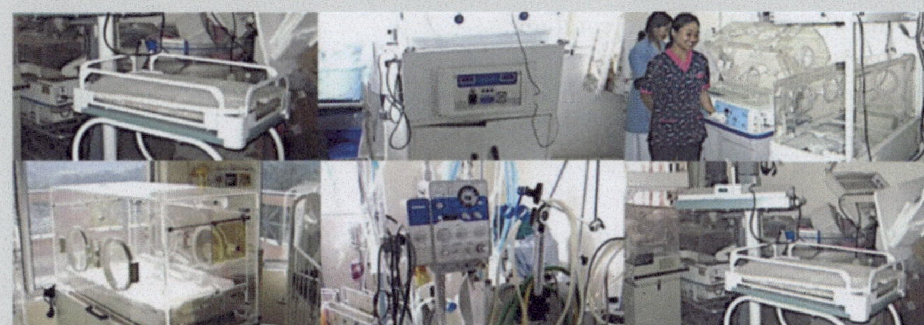

Empty incubators in India's regional hospitals

With these deep insights the team reframed the problem: "The need was not having a cheaper hospital incubator, but rather an inexpensive, portable and reusable baby-warming device, that could work without electricity, to be distributed to mothers and used at home. This marked a turning point in their design journey. The result is Embrace's Infant Warmer, successfully launched in India in 2011, a portable and an easy-to-use medical device that costs less than 0.5% of a traditional incubator. It consisting of an electric plastic-cased heater, a sealed pouch containing a wax-like phase change material (PCM) and a hypoallergenic sleeping bag that provides an acceptable temperature range so that a premature baby can be kept warm for hours without electricity.

(continued)

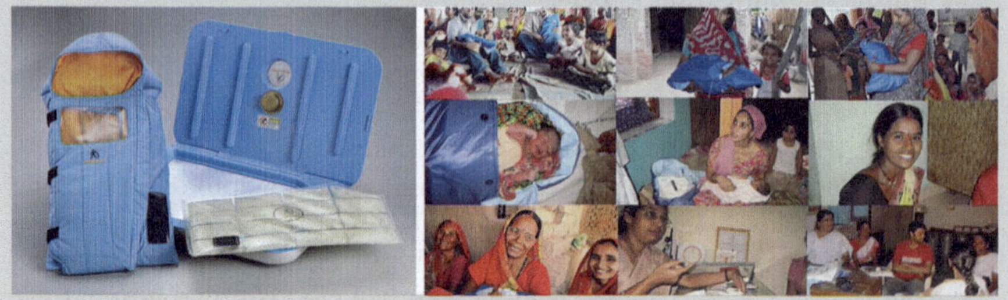

Embrace Infant warmer and first prototypes testing

The iterative process, based on understanding local needs, continuous insights and customer feedback allowed the team to reframe the problem and to depart from the traditional idea of incubator, and meet the real need of saving lives in distributed rural areas.

www.embraceglobal.org

www.changemakers.com/innovations4health/entries/embrace-infant-warmer

1.2.3 Divergent and Convergent Modes of Thinking

One of the key markers of an innovation process is the alternation between divergent modes and convergent modes of thinking, with frequent pivots between the two modes (Figure 1.6). Divergent modes involve generating multiple positions from a single point of departure, and convergent modes involve narrowing down to a smaller set of options or a more clarified, narrow point of view. The ability to switch from one to the other is called Pivot Thinking. It is important to note that many of our organizational and disciplinary processes are heavily biased towards convergent practices, and divergent modes cause great individual or organizational anxiety. The absence of this alternation between divergent and convergent modes implies that the practitioners hold a belief that at any time the single concept at hand is likely to be the right one. It is further indicative of the "creativity bias" whereby the author of a concept has a disproportionate level of belief in the worth of his idea.

1.3 The Steps in an Innovation Process

There are many innovation processes. The process and the tools dictate the modes of inquiry, and the types of solutions that emerge (whether they are products, services, human systems, technological systems, etc.). An innovation process

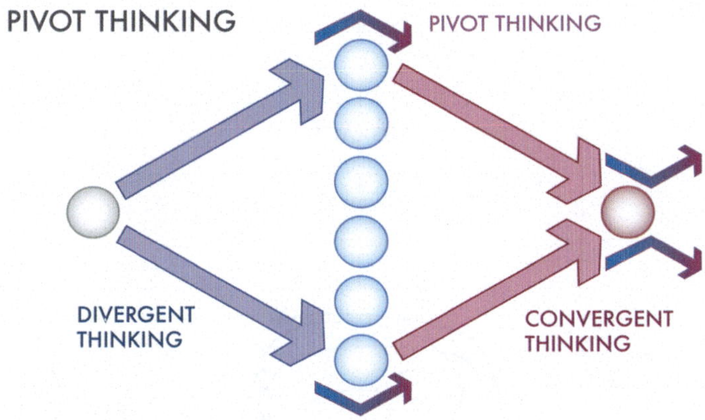

Fig. 1.6 Divergent and convergent modes of thinking

weaves between synthesis and analysis, and between abstract and concrete realms (Beckman and Barry 2007). An innovation process at its most fundamental level, involves a set of distinct steps that involve specific activities and cognitive modes. When people collectively go through these different cognitive modes together and innovate, it has an effect on the larger pattern of behavior at the system level and the organizational system starts operating in ways that constitutes innovation at a higher level. (Hutchins 1995). Figure 1.7 depicts the steps in a generalized innovation process. It is important to note that this is just one process, and there are many other innovation processes and tools. As long as genuine understanding and deep insights about the system are used to drive the design decisions, and creativity and prototyping are being leveraged in an iterative learning loop, innovation is taking place. The specific process might vary with problem type and the style adopted by a sub-community in the innovation world. For example, an innovator of physical products might follow a somewhat different process compared to someone designing an automobile.

The framework above shows the various stages of a human-centered innovation process. (Banerjee 2008) The discrete stages that are carried out iteratively and not necessarily in the same order are as follows:

(a) Deep immersion into the underlying causes of the problem.
(b) Exploration of the human dimensions of the problem, including unmet needs, motivations, mental models and cultural contexts.
(c) Use of synthesis as a cognitive mode to make sense of both qualitative and quantitative data in order to arrive at the key drivers and dimensions of the problem.
(d) Framing of the problems in creative ways, with a deep understanding of the genuine need as well as the contextual constraints and motivations.
(e) Making of strategic decisions, defining the boundary conditions, and crafting a theory of success to guide the solution.

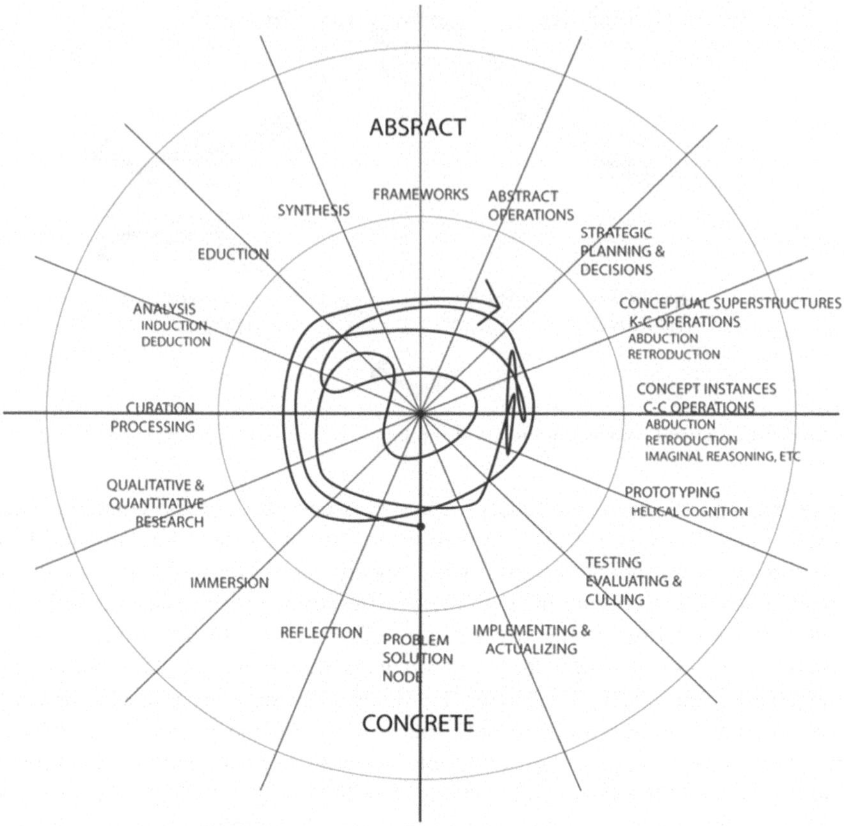

Fig. 1.7 The steps in a generalized innovation process

(f) Use of abduction (Takeda 1994) and creativity to generate multiple conceptual possibilities and populating of the solution space with many creative options. Abduction is the cognitive mode that involve imagining possibilities that do not exist yet.
(g) Use of rapid prototyping in order to simultaneously test creative options, but also as "probes" in order to understand the nuances of the problem and to eliminate failure modes early and inexpensively.
(h) Iterating on both the problem framing and the solution framing, and "multiplexing" between the abstract high level and the tangible low level.
(i) The development of the solution with a highly leveraged use of resources, including human resources.
(j) The act of representing and communicating the solution internally and externally (Suchman 1988), and creating a strong narrative and "story" that would help propel the proposition forward.

(k) Creation of a roadmap for implementation that continues to iterate on specific issues in order to eliminate failure modes before they get too expensive to fix.
(l) The creation of the right enabling systems to manage implementation and diffusion so that the process continues to address emergent and unforeseen issues, navigating towards scaled impact and system transformation.

The process of innovation is like a fractal – there is innovation needed in shaping the intentionality, and there is innovation required at the level of each tiny detail of a product feature. There is innovation involved in framing the problem, as in executing and scaling the solution. Since so many other processes in business, governance, and organization deal with all of these aspects, the "litmus test" lies in whether at any zoom level the process is designed to outperform normative modes. If a certain innovation methodology becomes the normative way, then it implies that new innovation methodologies need to be created to outperform those (Cross 2006). In this sense the entire game of innovation emblemmatic of a "growth mindset" (Dweck 2008).

1.4 Classes of Innovation Challenges

There are key distinctions in different types of innovation challenges. You can attain a tremendous level of expertise in one type of challenge, and be a novice in another. There are different classes of innovation challenges, and an innovation process that is well suited for one type of challenge might be completely unsuitable for another. Innovation capacity can be very high in a narrow set of problem types but low in others. Moving to new problem types without an understanding about its taxonomy can lead to being mistaken about one's capacity. It is important that we understand the class of innovation challenge we are undertaking, so that we can match it with the appropriate innovation techniques. The following is a framework that proposes a taxonomy of challenge types (Banerjee 2014, Fig. 1.8).

The figure above depicts three classes of challenges. The methodologies and pedagogy for Class A problems are better defined and well established. Class B challenges are understood in the some circles but most innovation communities lack expertise in the techniques. The methodologies, the ecosystem and the pedagogy for Class C challenges represent the frontier of the innovation field where the methodologies are undergoing rapid advancement. The following are different types of innovation challenges organized under the three categories.

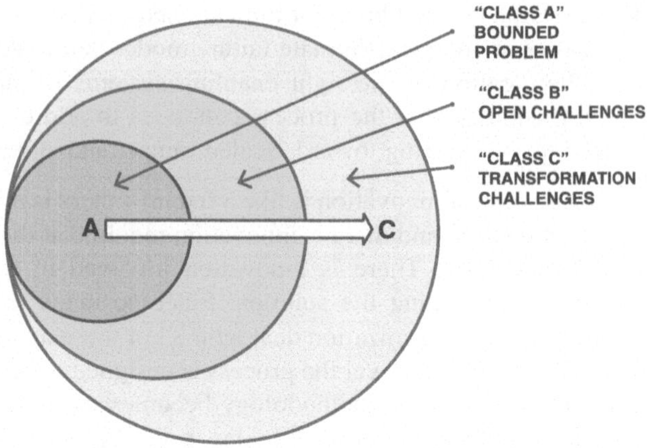

Fig. 1.8 Three classes of challenges

CLASS A: Implementation or Adaptation Type Problems

Type 1: Known and bounded problems with proven solutions. For example, developing an LED-based low-energy lighting solution that can replace incandescent bulbs.

Type 2: Open-ended but well understood problems with precedence of known solution archetypes. An example might be limiting traffic in the city center with expensive, zoned, limited-time parking.

Type 3: Semi-familiar problems with solutions adapted and extended from existing solution sets. For example, meeting the water challenge in developing countries with distributed installation of hand-cranked bore wells, water purification education, and water cooperatives.

CLASS B: Open Ended Medium Scale Design Challenges

Type 4: Pure technology or business or policy-based interventions that are directed at specific needs such as developing advanced battery technology for the solar industry, or a public health policy intervention related to subsidized health insurance.

Type 5: Closing system gaps through human-centered design and creative optimization. (Note: most systems are sub-optimal in their understanding of the human factor and deserve optimization.) For example, reducing the long waits at the Department of Motor Vehicles (the agency that issues drivers' licenses) through online services, advanced scheduling, automated form filling, and preemptive direct online billing. Another example might be designing the process of transition between nurse shifts in a hospital in a manner that prevents communication gaps that lead to incorrect medication and lapses in post-surgery care.

Type 6: Using Human Centered Design to carry out Experience Design or designing Pathways through a system (service design tends to make use of this approach) (Buchanan 1998). An example might be designing the entire "customer journey" for a public transport initiative that takes into account scenarios for bad weather, cultural and language barriers for visitors, picking children up from day-care, buying groceries on the way home and occasional non-dominant cases of journeys such as going to the airport with bags, meeting friends for a hike over the weekend, and returning home from a night shift in a factory located in a crime-ridden neighborhood.

CLASS C: Scaled Transformation Challenges

Type 7: Designing new system and scaling paradigms for unfamiliar and scaled problems. For example, reducing the energy footprint of an entire nation by a significant percentage through a combination of approaches, economic incentives, social innovation', and customized feedback.

Type 8: Transforming the behavior, roles, and relationships of the constituent stakeholders within the ecosystem, including non-human elements such as resource flows and natural systems. For example, an intervention to the food security problem by simultaneous engagement by banks, the government, the agro industry, farmers, small business enterprises, telecom companies, and non-profit companies.

Type 9: Transforming the behavior, outcomes and trajectories of the larger ecosystem. For example, changing the way in which our institutions, civic societies, industry, and government function, resulting in new direction regarding social, economic and environmental issues, and therefore building a more resilient, shock resistant future for our societies. Increasing the economic complexity (Hidalgo and Hausmann 2009) of an economy along with widespread change in behavior patterns, resulting in a large-scale change in the trajectory of economic development, would fall in this category.

Class A problems require innovation and design, but in bounded ways. Class B problems require tools that lead to deeper insights into the underlying human experiences and unmet needs, and design the user's journey through a system (Buchanan 1992). Class C problems require ecosystem level intervention. (Banerjee 2014) It makes use of all the methodologies from the other two classes, but require a more enhanced set of tools and disciplinary perspectives that can identify leverage points in a more nuanced and complex ecosystem design based intervention. (Meadows 1999)

The field of innovation is rapidly shifting towards challenge type C as businesses and institutions find it necessary to find leveraged ways of directing innovation towards more complex challenges. Institutions and organizations need to understand the taxonomy of different types of innovation challenges and create the capacity and expertise to use innovation processes for more complex challenges, and give leaders the ability to leverage and foster innovation. Organizations also need to understand the critical role that innovation plays in crafting their future, and need to foster forms of leadership that can bring about a culture of innovation.

1.5 A Departure from "Business as Usual"

Since innovation is defined by its ability to outperform "Business as Usual", it requires an ability to deviate from entrenched organizational norms, mental models, and decision-making styles that have come to shape "conventional wisdom":

1. **Understanding the Problem:** Often "Business as Usual" assumes that understanding of the problem and the anticipated objectives. In innovation processes the first step is to assume that you do not understand the real dynamics of the system and need to go about understanding underlying root causes.
2. **Empathy and Human Centered:** It seeks to understand the underlying human dynamics in the context of the challenge. As the human aspect is one of the trickiest layers, it requires the innovation team to be empathic and generate a deep level of insight about the underlying emotions, thoughts, cultural norms, and motivations. Often "Business as Usual" tends to be primarily focused on the technological or business aspects with very rudimentary understanding of real dynamics at the human level.
3. **Challenge and Outcome Centered:** "Business as Usual" typically introduces a disciplinary, technology, or domain expertise bias, whereas innovation processes are challenge centric and remain open about what type of solution might meet the needs.
4. **Needs-driven Solutions:** "Business as Usual" often drives solutions through specifications as proxies for the genuine need. Innovation processes focus on the genuine needs, and changes the definition of the solutions in terms of emerging understanding of needs.
5. **Willingness to Reframe the Problem:** "Business as Usual" allows entire institutional structures and processes to form around problem definitions that get cast in stone. Innovation methodologies rely on a tight feedback loop between new insights and perspectives and the framing of the problem. If the findings along the way contest the original problem statement, then the problem statement is considered to be flawed and ought to be reframed.
6. **Comfort with ambiguity:** "Business as Usual" tends to rush to clarify or converge as soon as there is ambiguity. Innovation processes embrace ambiguity and necessitates a much higher degree of comfort with ambiguity. Complex challenges require non-deterministic paths to understanding, because *"you don't know what you don't know"*. It requires dealing with partial knowledge, conjectures, synthesis, rapid theory building and testing, and an organic way in which the team understands the nuances of the challenge.
7. **Comfort with Failure:** "Business as Usual" abhors failure. Innovation processes acknowledge that there is no path to a robust solution without going through failure. Innovation methodologies *embrace* failure as a way to rapidly learn about failure modes early so as to design robust solutions.
8. **Reframing Risk:** "Business as Usual" abhors risk and tries to eliminate risk at every step. It also seeks evidence as a condition for decision, even if the problem has never been solved before. Innovation processes assume that risk-

averse behavior at every step is very risky, and mitigates the risk of innovative ideas through iterative prototyping and testing.

9. **Synthesis versus Analysis:** "Business as Usual" relies on *analysis* for most decisions. Innovation makes use of both analysis and *synthesis*. Analysis is the act of breaking down a problem into smaller parts and deriving confidence from how defensible a claim is. Synthesis is the act of joining dots, and seeing underlying or overarching patterns or threads. It involves joining seemingly disconnected facts to create meaning.
10. **Iterative Learning Loops:** "Business as Usual" has sequential decision chains, which make it very difficult to go back and change the definition of the problem while implementing a solution or strategy. Innovation works with an iterative "learning-framing-conceptualizing decision loop", where the decisions of the previous stages are revisited in the light of new information and emergent understanding.
11. **Rapid Prototyping as a way of thinking:** "Business as Usual" locks the problem definition before solving it and often gets to a fairly complete product or service architecture before testing. Innovation methodologies iteratively explore both the problem and the solution sides of the equation by using rapid prototyping as a way of thinking. Rapid prototyping is used to identify potential failure modes and the key forces at play.
12. **Divergent and Convergent Modes:** "Business as Usual" places little value on divergence, or diverges only in an incremental manner. Innovation relies on repeated oscillations between convergent and divergent modes with a keen judgment on when to pivot from one mode to the other. Divergence populates a space with multiple options, and convergence reduces a large number of options into a small set of "down-selected" ones. Alternating divergent and convergent modes creates a set of options that are a result of broader exploration and represents a higher state of clarity and understanding.
13. **Trans-Disciplinary:** "Business as Usual" places disproportionate faith in domain specificity and domain expertise, and sees breadth or crossing of silos as dilution. Innovation processes address a challenge in a trans-disciplinary manner, where a diverse set of disciplinary experts bring their collective points of view and transcend their disciplinary paradigms in order to create a *new thinking* that befits a complex challenge.
14. **Co-Creation:** "Business as Usual" generates ideas in mono-disciplinary silos within the context of a single disciplinary or organizational culture at any given time. Co-creation uses trans-disciplinary teams that *create together*, not unlike a jazz quartet improvising together.
15. **Multiple Cognitive Modalities:** "Business as Usual" often falls into a monological, reductive, deterministic way of thinking, without sensitivity to the need to switch cognitive modes. Innovation recognizes that there are many modes of thought and action and matches the mode of thought to the one that is optimal to use during a specific phase of the design process. For example an innovation team can go from being empathic, to being reflective, synthesizing, abstract framing, envisioning future states; being incisive and evaluative;

being generative, thinking through the act of building, dispassionately observing the results of testing, clarifying, and storytelling; or being a facilitator.
16. **Plurality:** "Business as Usual" believes that there is one right way or one right perspective, and sees any position outside the one that it believes in as false, or as a threat to one's position. Innovation methodologies take a more dualistic and pluralistic point of view, where multiple perspectives can coexist without necessarily competing with each other for existence. This also leads to a tolerance to multiple outcomes, goals, cultures, points of view, paths to success and definitions of success.

1.5.1 Risk and Failure

The issue of risk and failure deserves special attention. One of the key reasons that Silicon Valley succeeded in becoming a hotbed of innovation is its sophisticated understanding of risk and failure. Being risk averse is a basic human trait. We fear the unknown, the untested, and the unfamiliar. We have societal and cultural constructs around failure that are deeply ingrained in our identities and our habits of mind. In many cultures, if a person fails at an endeavor, they think of themselves as failures, and friends and family hastily sweep their story under the rug. Failure is seen as something that is to be ashamed of, disappointed with, and to avoid at all costs. Entire organizational processes and decision-making processes are crafted with extreme risk-averseness and a fear of failure.

In trying to eliminate risk at every stage, on one hand one is raising the threshold to failure modes, and on the other, simultaneously inadvertently lowering a ceiling for innovation. All genuinely innovative ideas are weeded out because they are considered risky and the only ones that are left are the ones that are considered "safe" and inherently incremental in nature. *Risk averseness at every step, results in a much bigger risk!* The most important implication of this is never building the capacity to carry out the kind of radical innovation and agility that is necessary to navigate complex challenges.

The fear of failure, while understandable, can be an extreme limiter of innovation. Firstly, a fearful state inhibits creativity. Playing safe is indicative of fear. In fearing failure, there is a tendency to play safe, make incremental moves, and not rock the boat. More importantly, trying to avoid failure is based on a theory that *there is a path to success without going through failure*. This is an egregiously narrow view of failure. Of the different types of failure there are typically only a handful of them that are to be avoided. The others are welcomed events that represent valuable opportunities to learn. If one can discover the failure modes early, then one can build a robust solution that has been designed to resist the failure modes. If those modes are not identified, they will eventually affect the initiative when it is much more expensive to change direction. In Silicon Valley, an entrepreneur who might have failed multiple times wears her entrepreneurial endeavors as a badge of honor and courage. The entire venture capital and entrepreneurial community in Silicon Valley understands the value of failure and the lessons that are learned from it.

Failure modes in bounded problems are easier to identify. For example, if you're designing a thin-walled casing for a small electronic device, you know that it would have to withstand a drop test and the material choice and geometry are developed with that specific drop test in mind. In more complex issues however, the understanding of the failure modes and the levels of risk, and the creative options for mitigating risk, are an *emergent property*. Given this, it is just understood that new failure modes and risk areas will emerge as a result of wading into complex challenges, and one would have to build the *capacity* in the process to continually address them. Instead of letting the fear of failure or risk inhibit innovation, it is more important to develop methods of mitigating risk of potentially path-breaking concepts.

1.6 The Need for an Innovation Ecosystem

Innovation is at once a discipline as well as a layer that can be added to any initiative to great effect (Jucevicius & Grumadaite 2014). We have discussed how building innovation capacity is a tremendous exigency for any institution in its effort to weather the stormy seas of the future. Innovation is often antithetical to the cultures of institutions that are in dire need of transformation to meet rapidly changing external contexts. An institution could be extremely satisfied with its level of innovation because it is making incremental advancements in a narrow discipline-centric silo, and completely lack the vision and drive for creating conditions that allows breakthrough trans-disciplinary innovation.

Whether innovation capacity is at the level of an individual or an institution that wishes to create the capacity internally, it can only flower in an ecosystem that fosters innovation. The ecosystem can be created in a bottom-up manner or in a top-down manner, but since the value systems of innovation are often at odds with current practices, a certain engagement and commitment on the part of the institutional leadership is necessary for an innovation ecosystem to be instantiated, take root, and become a pervasive part of the culture.

References

Banerjee Banny (2009) Design: A field in Transition, Innovation Journal.Innovation, Journal of the Industrial Design Society of America, Winter 2008, Vol. 27 No. 4, 18

Banerjee Banny (2014) Innovating Large Scale Transformations, book chapter, "Design for Policy" edited by Christian Bason, Ashgate Publishing Ltd.Company, Burlington, VT, USA, 71–86

Banerjee Banny (2008) Designer as Agent of Change: A Vision for Catalyzing Rapid Change, Changing the Change Conference Proceedings.

Beckman SL, Barry M (2007) Innovation as a Learning Process: Embedding Design Thinking, Harvard Business Review, November 01, 2007

Boden M (2004) The Creative Mind: Myths and Mechanisms, Routledge, New York, NY: 11–24
Buchanan R (1992) Design Issues, Vol. 8, No. 2 (Spring, 1992), The MIT Press, 5–21
Buchanan R (1998) Education and Professional Practice in Design" in Design Issues 14:2: 63–66.
Cross N (2006) Designerly Ways of Knowing, Springer-Verlag London Limited, 77–92
Deleuze G, Guattari, F (2004) A thousand plateaus: Capitalism and schizophrenia. London: Continuum
Dorst K, Cross N (2001) Creativity in the design process: Co-evolution of problem–solution. Design Studies 22(5):425–437
Dweck CS (2008) Mindset: The New Psychology of Success. Random House, New York: 124–133
Frayling C (1991) Research in the Art and Design, Royal College of Art Papers
Hatchuel A, Weil B (2003) A New Approach to Innovative Design: An Introduction to C-K Theory. International Conference on Engineering Design ICED 03, Stockholm, 5–13
Hayek FA (1964) The Theory of Complex Phenomena, Chapter 22, The Critical Approach to Science and Philosophy, edited by Mario Bunge, Collier-Macmillan Limited, London
Hidalgo CA and Hausmann R (2009) The building blocks of economic complexity, Proc Natl Acad Sci U S A. 2009 June 30; 106(26): 10570–10575. Published online 2009 June 22. doi: 10.1073/pnas.0900943106
Hutchins E (1995) Cognition in the Wild, MIT Press, Cambridge, 287–352
Jucevicius G, Grumadaite K (2014) Smart development of innovation ecosystem Procedia – Social and Behavioral Sciences 156:125–129
Levin K, Cashore B, Bernstein S, Auld G (2012) Overcoming the tragedy of super wicked problems: constraining our future selves to ameliorate global climate change. Policy Science, Springer Online 45:123–152
Meadows, Donella (1999) Leverage Points, Places to Intervene in a System, Sustainability Institute, VT, 1–19
Rittel, Horst W. J., and Melvin M. Webber (1973) Dilemmas in a general theory of planning, Policy Sciences 4, no. 2
Simon HA (2000) Keynote Speech, 2000 IDPT Conference.
Simon HA (1968) The Sciences of the Artificial, M.I.T. Press, Cambridge
Suchman LA (1988) Representing Practice in Cognitive Science. Human Studies 11:305–325
Takeda H (1994) Abduction for Design, Proceedings of the IFIP TC5/WG5. 2 Workshop on Formal Design.
Thackara J (2005) In the Bubble, Designing in a Complex World. MIT Press, Cambridge, MA
Webber (1973) Dilemmas in a general theory of planning, Policy Sciences, Volume 4, Number 2 (1973), 155–169, DOI: 10.1007/BF01405730

Chapter 2
The Innovation Ecosystem

John Body and Fawwaz Habbal

2.1 Why an Innovation Ecosystem?

The big issues we face today are those that go to the heart of our societies, economies and environments. What is the best response to terrorism? What is the right approach to immigration to enhance society and build the economy? How can we preserve water resources while achieving economic returns from irrigated agriculture in a way that fits with the local context? These are just a few examples of the intractable challenges that we face today. These are the types of problems that C. West Churchman called "wicked problems" in his "Guest Editorial" of *Management Science*. Wicked problems resist resolution and have many interrelated parts such that a change in one area will affect the other parts, often in an unintended way (see also: Rittel and Webber 1973).

The innovation ecosystem mobilizes around wicked problems, sometimes at a local level and sometimes at a national or even global level. The innovation ecosystem is the place to bring social, economic and environmental problems for attention. Because of the inherently complex nature of wicked problems it is not possible for one person to hold all the expertise and knowledge to attenuate those problems. It requires the many and varied disciplines involved to come together to work in ways that transcend the individual disciplines. For example, a complex health issue could require the perspectives of medical practitioners, medical educators, service providers to the health industry, providers of medical software systems, accreditation and standards setting organizations, medical ethicists,

J. Body (✉)
Founding Partner, ThinkPlace, PO Box 5249, Kingston, ACT, Australia
e-mail: John.Body@thinkplace.com.au

F. Habbal
Harvard School of Engineering and Applied Sciences, Harvard University, 29 Oxford Street, Cambridge, MA 02138, USA

B. Banerjee, S. Ceri (eds.), *Creating Innovation Leaders*,
Understanding Innovation, DOI 10.1007/978-3-319-20520-5_2

privacy advocates and guardians, government health agencies, central government fiscal agencies and, of course, the health consumer.

The innovation ecosystem acknowledges that wicked problems need non-normative approaches. There is no formulaic approach to develop non-normative solutions. To say there was a set formula would be an oxymoron because then it would be a normative approach. However, the innovation ecosystem can suggest some elements, qualities and relationships that will increase the likelihood that a wicked problem can be addressed.

The innovation ecosystem described in this chapter includes the components, the connections between those components and the culture that together comprise the system.

2.1.1 *An Overview of the Ecosystem of Innovation*

An ecosystem of innovation results from the complex intricacies between pedagogy, practical processes, delivery systems and services. The innovation ecosystem is working well when educational institutions, strongly endorsing innovation leadership, with partner stakeholders, such as government agencies, industry, start-ups, venture capitalists, and non-profit organizations, work together to build new types of platforms and relationships that create impactful outcomes [an early example is the renascence area and the Medici; see Johnasson 2006]. Supplementing the tangible parts of the ecosystem are intangible parts, importantly a culture that fosters innovation, entrepreneurship, embracing complex challenges and new thinking. These components and cultures combine around emerging issues, such as social, economic or environmental issues, to increase the likelihood of making advances around wicked problems (Fig. 2.1).

In academic institutions heavy emphasis is placed on mastering and developing depth of subject-specific expertise in clear domains. By contrast, a knowledge base where the expertise is about bringing different disciplines together and overlapping domains is less valued. Yet, ironically, it is in bringing together diverse perspectives that new knowledge is created. The 1997 Harvard Business Review article "Putting Your Company's Whole Brain to Work" by Dorothy Leonard and Susaan Straus talked of the concept of "creative abrasion". Whilst the concept of innovation is well accepted, Leonard and Susaan argue that achieving innovation is hard. It is hard because innovation occurs when different ideas, ways of knowing, ways of

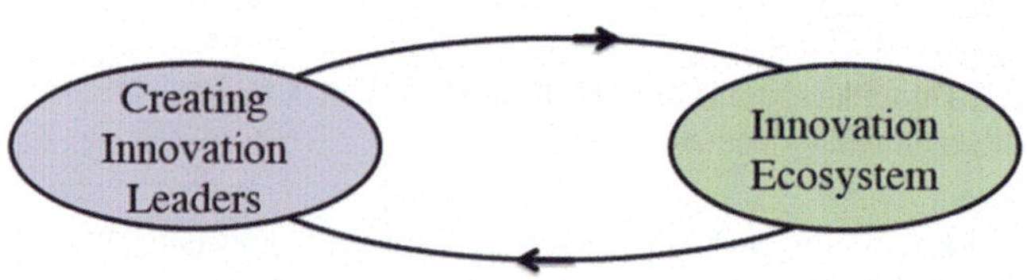

Fig. 2.1 An interplay between creating innovation leaders and the innovation ecosystem

processing and ways of judging combine. And leadership does not know how to bring these differences together in a constructive way. Instead of constructive innovation these differences often result in combative behaviors that have the effect of stifling fragile new concepts.

That was 1997. Since then small advances have been made in the way disciplines work together; yet the foundational issues remain. Academia, industry and government still do not value these types of trans-disciplinary approaches. Experts in trans-disciplinary ways receive less recognition than their counterparts with deep domain knowledge. To change this, efforts are required to build the supply chain and the training grounds for innovation leaders. In addition, significant efforts are also needed to advocate for changing the mind-sets and cultures of potential employers to create jobs at all levels so that innovation is honored and rewarded.

Arguably every organization invests in innovation, however small that percentage of investment may be. With no capacity to innovate, an organization has no capacity to respond to the enormous tides of change that are evident globally. Not investing in innovation will result in the rapid or slow demise of an organization, depending on the speed at which their environment is changing. There are many examples where companies that have been at the top of their field can rapidly drop from industry leader to a second, third or fourth place because of a competitor's innovation. Government agencies may feel less prone to competition; however, a government agency that fails to respond to its environment will progressively see budget reductions, forced restructures and amalgamations as a result of their non-performance or progressive irrelevance.

Evidence of the susceptibility of businesses can be seen by comparing the changes in stock market leaders over five-year increments. Equally, susceptibility of government agencies to change can be made by making the same comparison over time. Newly elected governments make changes in response to their policy stance and in relation to past performance.

Some organizations invest heavily in innovation, such as those involved in technology. Other institutional organizations may invest a smaller proportion but they are nonetheless investing in new technologies to improve the way they reach customers and to improve the efficiency with which they perform their part of the supply chain. This is true of any organization: universities, commercial organizations, government agencies and non-government organizations.

When describing the innovation ecosystem, there is a range of characteristics that can be helpful. First, there are the ***entities*** that make up the ecosystem, the academic institutions, the governments, the enterprises (established and start-up) and the individuals that make up society. Second, there are the **relationships** between the entities that are a key ingredient to innovation. Third, there is the ***innovation cycle*** that provides reinforcing learning, progressively enhancing the ecosystem's ability to address complex innovation. Fourth, there are the ***innovation layers*** that provide the opportunity areas where innovation can be applied. Finally, there are the values and behaviors that set the ***innovation culture***. These are the intangible characteristics of the ecosystem that have an equally significant impact on the ecosystem health to that of the tangible characteristics.

THINKPLACE

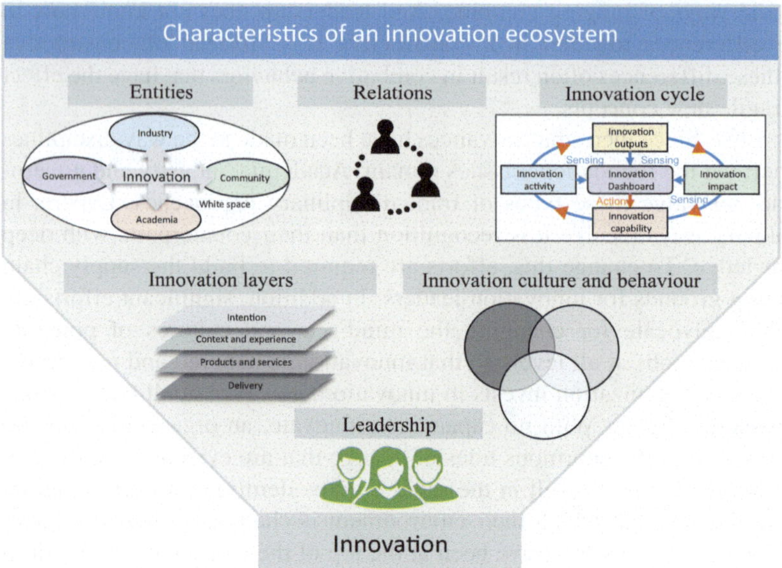

Fig. 2.2 Characteristics of the innovation ecosystem

The diagram below shows these characteristics of the innovation ecosystem (Fig. 2.2).

2.1.2 Entities That Make up the Ecosystem

The four broad clusters of entities in the ecosystem are shown below. All of these types of entities play an important role in shaping the ecosystem (Fig. 2.3).

Academia, through universities and other educational institutions, carries out innovation research; they develop curricula and, in turn, through delivering those curricula, prepares future leaders who will sponsor innovation in the contexts they find themselves in. Within academia we see students, researchers, teaching staff, curricula, university leaders, intra-disciplinary experts, trans-disciplinary experts as well as incubators and start-ups that emerge from universities.

Governments shape the appetite of a country or state for innovation. Their policies and strategies can shape innovation incentives. They shape and fund educational systems. They can provide grants and tax incentives to sponsor research. They set performance measures for universities and through that can shape innovation. Governments are also in the business of innovating their delivery models as they deliver services to the community. Within a government there are those who influence education policy, those who influence industry and innovation policy and law, those who set regulations and deregulations, those who broker relationships between parties in the ecosystem and those who allocate grant funding.

Fig. 2.3 Types of entities play an important role in shaping the ecosystem

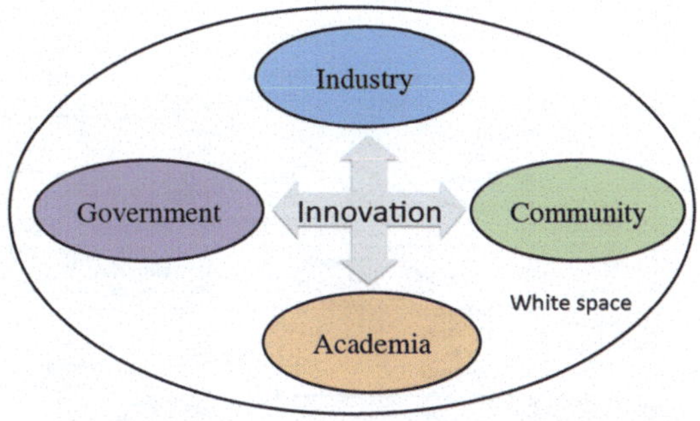

Industry generates the wealth of a country or state. To a large extent the sustainability of that wealth creation is dependent on the organization's ability to innovate new products, services and delivery models. Innovations could be in a new product or service, in an improved one or in streamlined and more efficient delivery models. Within industry, it is not just the traditional publicly listed companies that form the innovation ecosystem. Equally important are the start-up companies, those providing seed funding, the social and commercial entrepreneurs, the venture capitalists and those brokering and advising other parties. In developing contexts, the Non-Government Organization plays a key role, often sponsoring programs that in a more developed context would be the domain of government. Some common territory for innovation in the non-government sector includes health, agriculture, food security and finance.

The **community** is another key player in innovation. The community may have a higher or lower appetite for innovation. This can set government's and industry's appetite for reform as well. Enabled by social media the community can much more readily share its ideas and concerns with others. And, increasingly, communities are mobilizing around ideas and initiating a groundswell of innovation in a self-organized way.

Finally and most importantly, there is the innovation pipeline that plays out in so many different contexts, with different parts of the ecosystem involved, whose function is to take ideas and develop these in ways that create value. Value is increasingly being looked at not just as value to the organization but shared value across a greater range of stakeholders.

The four core components described above are expounded upon below. The diagram illustrates that there are many parts that have to come together for innovation. Within each of the broad categories there are key components that must be present (Fig. 2.4).

2.1.3 Relationships That Foster Innovation

The innovation ecosystem is as much the connections as it is the components. The connections are pivotal to innovation because they allow transformational ideas to

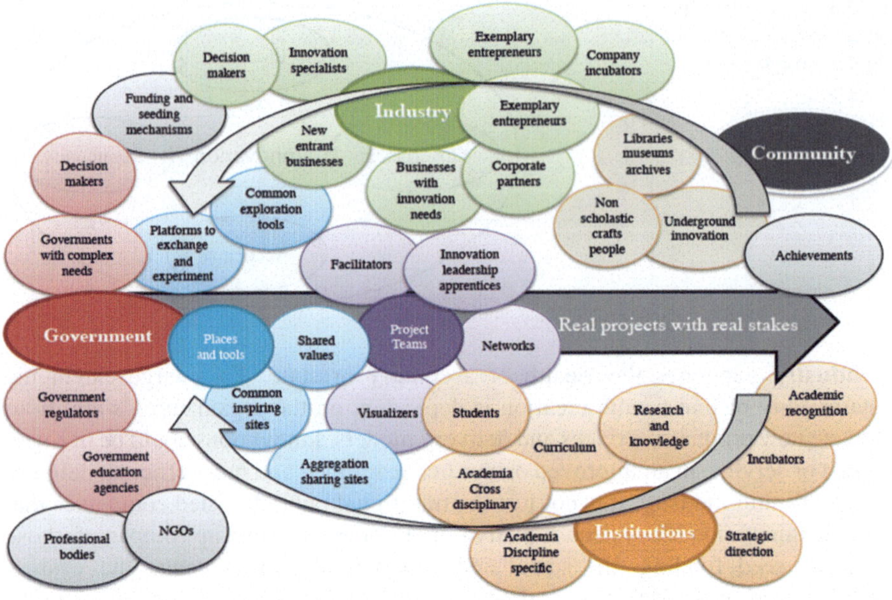

Fig. 2.4 Innovation Leadership ecosystem

be generated. For example, a start-up entrepreneur may receive a grant from government, or seed funding from an investor. Its concept may be part of a value chain that requires connections with others in the value chain.

Evidence of relationships within an innovation ecosystem includes networks of connections, positive and constructive relationships combined with collaboration and co-innovation.

A strong innovation ecosystem will generally be observable through deep connections, for example from customers, within the supply chain, to regulators, to universities and to funding sources.

2.1.4 The Innovation Cycle

The cycle in a healthy innovation ecosystem is self-generating, progressively building the innovation capability through reflective practice. In a healthy ecosystem the innovation capability is growing as the result of reinforcing feedback in the system. A healthy innovation ecosystem will progressively learn and add to explicit and tacit knowledge and expertise. We see this learning behavior in academia, industry and governments and it is an indication of a healthy ecosystem.

At the bottom of the cycle is the innovation capability. Innovation activity draws on the capability to generate innovation outputs. These outputs in turn have social, economic or environmental impact for the individual or the collective. A healthy

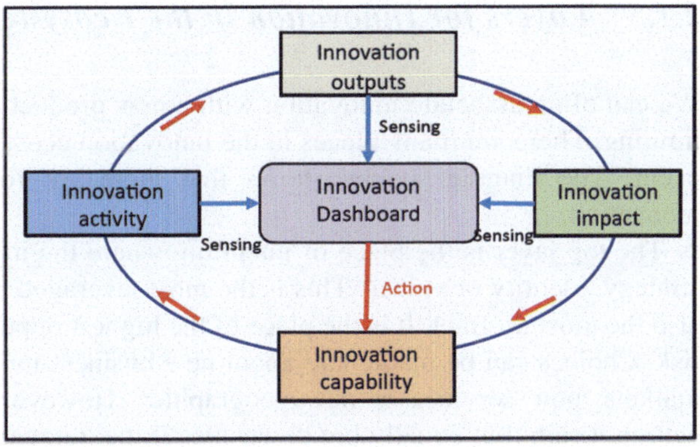

Fig. 2.5 The innovation cycle

innovation system builds knowledge and capability that over time enhances its ability to innovate further (Fig. 2.5).

Innovation capability can be codified into processes, systems and methods to become explicit capability. It can equally be part of individuals' and communities' knowledge and expertise, becoming tacit capability that can be drawn on. The center of the cycle shows the assurance mechanism that monitors and measures the ecosystem, allowing future iterations to respond to more complex challenges.

Mobilizing the ecosystem requires that the system have the ability to identify the need and challenge, define this as a problem, and then generate innovations. This is at the heart of the innovation activity in the ecosystem. The innovation activity identifies the boundary conditions of a situation or issue and the dimensions along which there is a need to change it. Another way to look at this is the gap between existing and potentially better states, even if the desired future state has yet to be framed and articulated.

Once a challenge has undergone the process of root cause analysis, problem identification, and framing, then the situation or issue has been "problematized", and lends itself to "problem-solving". Often the bigger value lies in identifying what the problem is rather than simply solving for a pre-identified problem.

A clear problem framing is the precondition to generating innovation. Innovations are non-normative solutions that outperform normative solutions. They typically involve creativity, based on insights that provide new ways of understanding. The innovation takes account of the multiple criteria for success.

The innovation activity generates innovation outputs leading to impacts. A well-crafted innovation should maximize positive impacts and minimize through anticipation unintended negative impacts. Evaluation occurs at the center of the cycle. Measuring the individual projects and the collective ecosystem is the topic of the next section, which discusses measurement of the ecosystem.

2.1.5 Layers for Innovation in the Ecosystem

We can often associate innovation with a new product or service. But this view is limiting. There are many places in the innovation ecosystem where innovation can occur. The diagram below shows that there are four layers to consider for innovation.

The top layer is the place of intention, where the innovations can be in policy, strategy, identity or values. This is the most leveraged place of innovation but it is also the most abstract. It is the place of the highest opportunity and also the highest risk. Choices can be made here about new business models, new positioning, new markets, new services or new geographies. However, there are big upsides to getting it right but equally big downsides if the innovation in strategy is wrong.

The second layer is gaining increasing interest as a space for breakthrough innovations. This is the reality space where people or groups interact with their context and organizations. This is the reality of lived experience. Increasingly universities, industry and governments are seeing the value of intentional innovation in this area. For too long people have had to make sense of poor quality interactions with institutions that may have unintentionally gravitated to solutions that worked for the organization rather than the customer or citizen.

The third innovation space is in the area of products and services. This is the traditional domain of innovation, where new products and services are researched and developed through to production and implementation. It is worth noting that this third layer does not sit by itself, but rather is an integral part of the other three layers. This means products or services should be designed recognizing the organization's strategy and positioning, and the desired experience for customers and with consideration for how the product or service will be delivered.

The fourth layer appropriately sits at the foundation. It is the delivery layer. This is the layer that organizations can mistakenly over-focus on, with the organization becoming an end in itself, rather than understanding itself in context. This is because there is a lot of work to be done in the foundational layer. However, too much focus here can divert the organization's attention from the fact that there is a customer with a need to be met and this is where all the internal work should be directed. The foundation layer includes any internal levers that can be moved to affect the customer experience and the needs of other stakeholders.

Typical levers include the culture, skills and deployment of the workforce, the organizational structures, the business processes of the organization, the technology and systems and the budget allocations. This fourth layer offers considerable territory for innovation. Note that the fourth layer could cover more than one organization. Increasingly governments, industries and the non-government sector are working together on breakthrough innovations. This is particularly so in the developing world context.

In the top layer policy, strategy and brand are generated.

In the second layer, the innovation process zooms between the deep-lived experience and the whole system, ensuring both are viable and congruent.

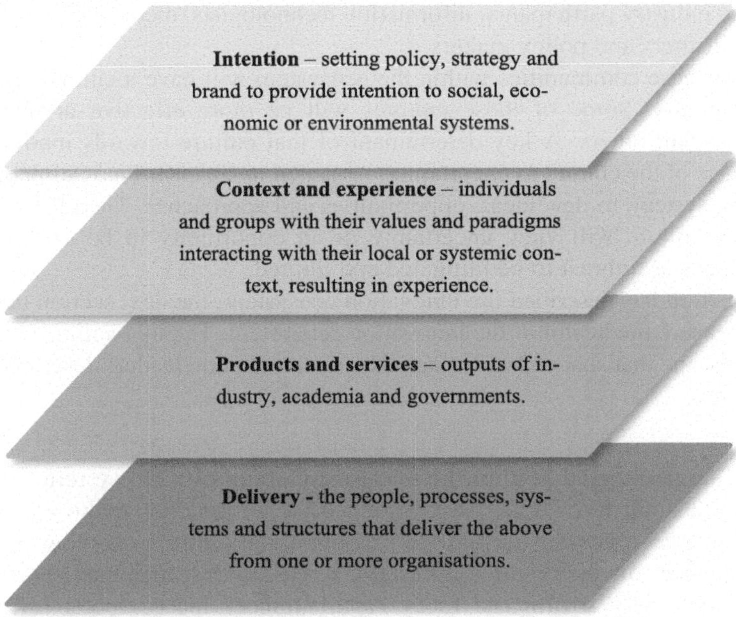

Fig. 2.6 Multiple layers in complex systems providing spaces for innovation

The third layer provides the innovation space for the components that shape human behaviors and experiences.

The foundation layer is the platform where innovations can occur within and between organizations and entities (Fig. 2.6).

2.1.6 Innovation Cultures and Behaviors

Whilst there are tangible elements of the innovation ecosystem such as the entities, relationships and cycles of innovation, there are important intangible elements.

Culture is intangible yet very real because it shapes the behaviors and decisions that people make on a daily basis. Culture has to do with ways of thinking and the consequent actions and interactions that result. Culture is built from a set of beliefs and customs of the people in that community.

From an innovation ecosystem perspective, the culture is contextual to the community that forms the part of the ecosystem. That community culture could be defined by an institution, for example the innovation ecosystem surrounding Stanford University. Or it could be defined by a geography, for example the innovation ecosystem of funders, NGOs, scaling partners and start-ups in Eastern Africa. Or the culture could defined by a topic such as the innovation ecosystem surrounding genome mapping globally, including researchers, institutions, diagnostic industry participants, information technologists, medical ethicists, legal advisers, insurers and policy makers.

Each of these communities within the ecosystem will have a culture unique to that community. Some of those cultures will be more effective at incubating innovation than others. A key determinant of that culture towards innovation is the response of the culture to uncertainty. As stated in Chap. 1, innovation operates in an open system, to new ideas, opportunities and approaches. Therefore a strong innovation culture will view uncertainty as an opportunity to be explored and exploited, not as a threat to be mitigated and limited.

This section has described the innovation ecosystem; the next section describes how to measure the health of the innovation ecosystem. The innovation ecosystem is the landscape that shapes and is shaped by the innovation leaders described in the next chapter.

How Google Started Within the Stanford Community Ecosystem

Google has transformed the way we search for information, making it 'universally accessible and useful'. Its rule-breaking search technology is the consequence of the genius, foresight, focus and leadership of its of co-founders Sergey Brin and Larry Page, engineer and computer scientist who met at Stanford University, and of the strong support of the Stanford University ecosystem they were placed into.

Interested with links on Web pages in 1996 they started working together, with the idea that, as with logic of book citations, links from important pages are more important than links from pages that are less important. Based on this intuition, they developed an innovative link-rating system, "PageRank", that calculated the relevance of a web page to the user's query based on the number of other pages that linked to it, making sense of the vast heterogeneity of the World Wide Web. They tested the technology that would become the heart of Google, launching a search engine for the Stanford community internally accessible by students, faculty and administrators at google.stanford.edu that quickly grew its popularity. PageRank technology was incubated in the Office of Technology Licensing as part of a portfolio of promising Stanford technologies; a patent for PageRank was acquired and several Internet companies were contacted.

Computer science professor David Cheriton, introduced them to Andy Bechtolsheim, founder of Sun Microsystems and vice president of Cisco Systems, a technology company in Silicon Valley (Bechtolsheim, Cheriton and Bezos are Amazon's founders). David was impressed by the way Google solved the simple problem of finding information and was impressed by the

(continued)

intellect and drive of the two young inventors and after a brief discussion Bechtolsheim wrote out a check to Google Inc. for $100,000 without knowing that the company didn't even exist yet. In 1998, they launched Google with an index of about 60 million pages with search results that were better than those of competitors like Hotbot or Excite. Shortly after that, Brin and Page could enlarge the collection of Google-wares; they added eight phone lines, a cable modem and a DSL line and rented a garage in Menlo Park in Palo Alto with a sign saying "Google Worldwide Headquarters".

Brin and Page envisioned long-term, deeply innovative and highly scalable solutions with the help of the Stanford ecosystem and the strong power of its network, active in fostering innovation.

The Stanford human ecosystem where Sergey Brin was a student, with excellent people (Jim Gray, Moshe Vardi, Jeff Ullman, Hector Garcia-Molina, Jennifer Widom, Alon Halevy, Alberto Mendelzon, David Maier, Rakesh Agrawal, and many others) and Sergey Brin and Larry Page in their first Google headquarters at 232 Santa Margarita, Menlo Park, California

Further readings
Scholes (2000), Vise and Malseed (2005), Brin and Page (1998)

An Exemplar Case: The Music Ecosystem
In an era of pervasive media, music became ubiquitous, consumed 'every time, anytime', while concepts such as ownership and acquisition slowly became increasingly irrelevant. Instead of thinking "How do we make people buy more and more music?", new market players radically looked beyond the traditional boundaries, including online and webcast radio, video platforms, subscription music services, and virtual stores, adopting a strategy whose key aspect was not to sell products but rather services and access to music, with new pricing formulae and reshaping in a few years the entire music ecosystem

(continued)

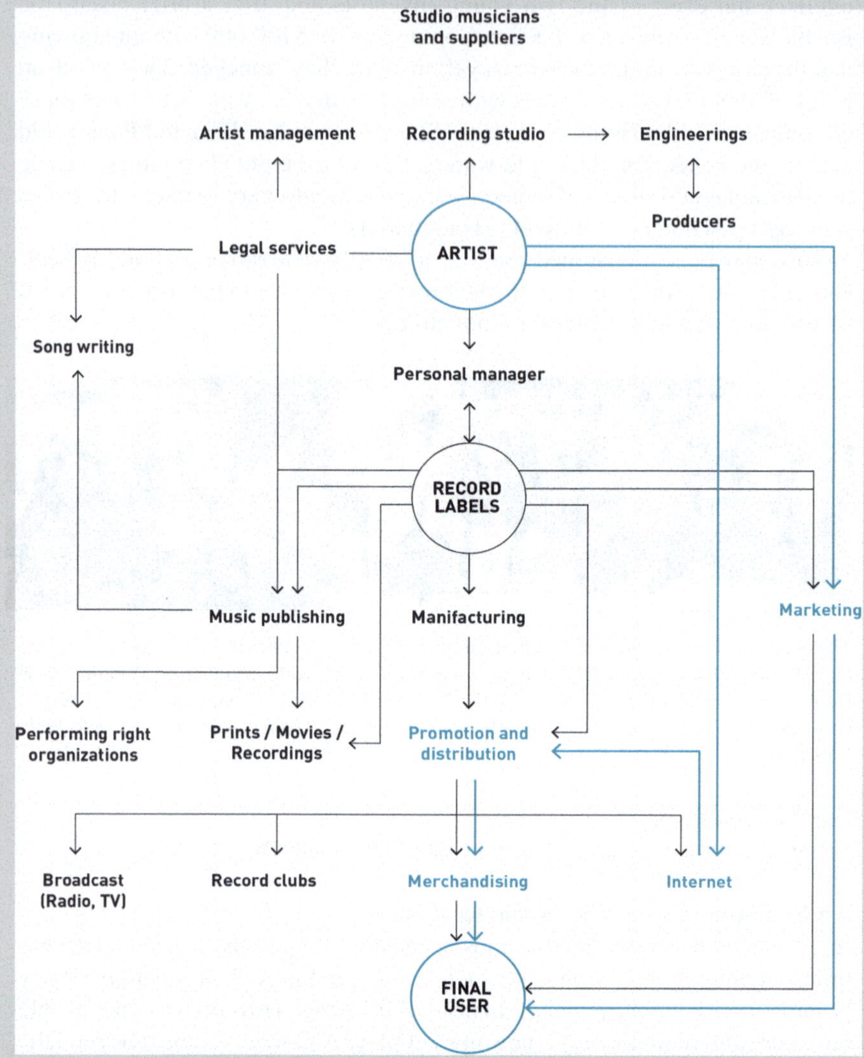

Changes in the record labels-controlled music value network. A new ecosystem based on record labels disintermediation (blue)

As shown above, in the new music value network marketing and promotion have partly been migrated to the Internet. Audiences have a much more active role in sharing, uploading, commenting and remixing. Manufacturing

(continued)

and distribution have partly been replaced by digital technologies for home recording and file sharing. Distribution cost decreases contribute to flattening the tail by making it easier for niche products to enter the market and foster the development of decentralized promotion: music artists can now directly link to their audience, without interference of a music label. Innovation into new markets has been possible through the disruption of the value chain by new innovation leaders, who have turned out to be the new key actors of the whole music ecosystem.

2.2 Measuring the Innovation Ecosystem

2.2.1 Measurement as Part of the Innovation Ecosystem

Innovation emerged as a lifeboat in turbulent sea of dramatic challenges of the 21st century, be it the scarcity of resources, the shifts in the cultural and social norms or the loss of the commercial boundaries and the rise of globalization. Innovation is critical as it is a challenge centric process that transcends disciplines, with the aim to create processes and outcomes to outperform a system's normative outcome. By mobilizing the system capabilities, innovation focuses on generating new perspectives, transformational ideas and flawless execution.

Leaders of innovations must have a full awareness of the health of their ecosystem, and in fact they must feel responsibility for monitoring the health of the ecosystem. This is not a trivial task and requires the presence of organization and infrastructure. For example, the presence of explicit elements of processes, practices, technologies, methods and techniques that enhance people awareness of the importance of innovation, what innovation is, and how to contribute to innovation and increase the innovation capability and velocity are necessary for the health of the organization and its infrastructure. Innovation leaders should be instrumental in installing such infrastructure. It is important to note that such infrastructure cannot be imposed on the ecosystem; rather it must an integral part of it.

To be able to direct the ecosystem towards positive and higher states, innovation leaders use tools to measure elements that determine the direction, velocity and capacity of the innovation cycle. Thus measurement has a very special importance to the execution and effectiveness of innovation. Leaders of innovation must be in aware of the different measurements, understand their value and limitations.

2.2.2 *Measurement as a Strategic Tool*

Innovation changes the ecosystem, and also with the feedback loop the ecosystem changes innovation, its tools, processes, and its outcome. This dynamics requires an embedded measurement practice. Innovation leaders, monitoring the feedback loop, are able to direct and adjust the course of innovation (Fig. 2.7).

Part of the innovation process might be to evolve and change known elements such as utility of a process, adapt and improve particular boundaries, or reframe an offering. However, transformational innovation may remove boundaries and bring about unknown outcomes to unfamiliar challenges as part of the quest to create an outperforming value. Regardless of the degree of outreaching output, measurements are required not only to evaluate the return of the investment and preempt poor outcomes, but also to know the efficacy as well as to create expertise. In some sense measurement is an early indicator of the future state and a feedback tool to modify and possibly alter the future state. In addition, measurement is important to build theory and test models, and since innovation is a response to a cause and aims at creating a change, measurement provides a unique tool to amplify the impact.

Unlike the teachings of an exact science, where measurement is a fact finding or evidence-seeking tool, creating measurement for the forward looking innovation is not easy and requires understanding proxy. In addition, sensing and making sense of an innovation attribution is as important as a quantitative or qualitative measurement. Innovative leaders develop an inner feeling for 'innovation sensing' and become experts in navigating the innovation ecosystem to successful outcomes, in a spontaneous fashion similar to a car driver sensing the road conditions and then making the proper driving decisions.

An important side effect may happen as measurement is taking place. Normally, the measuring tools are part of the ecosystem, and it is then possible for the action of measurement to interfere with the dynamics of the innovation. This internal feedback may only lead to inaccurate measurement, but it is also possible that may cause a better alignment of the objective of the innovation action.

Fig. 2.7 Measurement is a tool to increase probability of success and direct innovation to attain the desired state

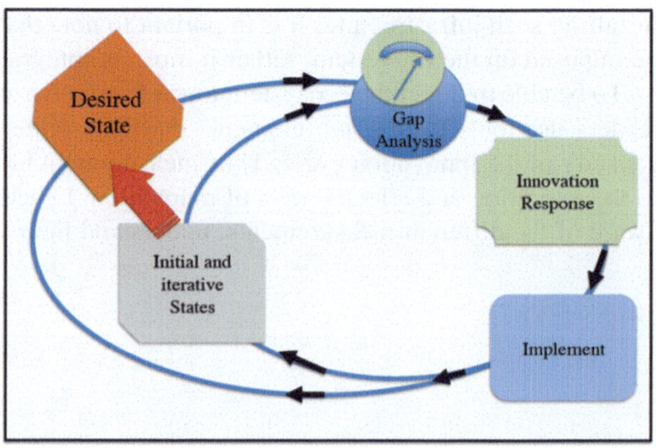

2.2.3 *Measurement in an Ecosystem of Several Interacting Subsystems*

Most human ecosystem contains interacting multi-dimensions that cannot be represented by linear measurable states. When innovation is exercised in any of these dimensions, complexity creates dynamics that require careful tracking. Innovation is not an action to improve an independent single outcome such as the cases for product offering or some process improvement, and what can be perceived as a positive innovation action may have other consequences that could be positive, or may cause issues that cannot be tolerated. Innovation outcomes may create shifts in parts of the ecosystems that could also cascade into other parts. Therefore the collective impact of these innovation outputs must be assessed, collectively and individually, to determine whether they are creating value to society, value to the economy (organizational efficiency or overall economy) and value to the environment (local environments, the global environment, the built environment and the physical environment). For example, in a given ecosystem an innovation element might be embedded in configurations for creating financial value, networks and connections, organizational structure and delivery processes, and customers' engagements such as service, delivery channels, and brand identity. An improvement leading to an enhanced financial value in a part of the ecosystem might create indirect poor consequences on service or delivery, and such negative consequences may make this innovation socially unacceptable. On the other hand, an innovation in delivery might create indirect financial gain that increase tax income and provide more flexible income for schools.

To illustrate the general concept, we consider the state of the ecosystem (A) and its associated interacting systems, or subsystems, each having a number of interacting elements (n). We call these elements $\{A_1, A_2, A_3, \ldots, A_n\}$.

Innovation $\{I_1, I_2, I_3, \ldots, I_n\}$ may need to be introduced to improve each one of these elements, or some of them. Thus a new state (B) will emerge with new elements $\{B_1, B_2, B_3, \ldots, B_n\}$[1]. These new elements $\{B_i\}$, some of which might the same as the old elements, may have new dynamics leading to new type of interactions in the new state (B). Given that, and the fact that it may not be possible to create measurement and evaluation for all the elements of (B), one has to make careful choices for what to measure, and design measurement techniques to obtain information through specific indicators, keeping in mind that some of these indicators might be quantitative and others require qualitative scale.

To illustrate this complexity, we show below innovation actions $\{I_i\}$ transforming elements $\{A_i\}$ to new elements $\{B_i\}$ through a *convolution* [illustrated in the following Fig. 2.8 by (X)] of $\{A_i\}$ and $\{I_i\}$. The output $\{B_i\}$ is then unpredictable and it is not the accumulative independent outputs of each innovation action.

[1] The output of innovation can be viewed as a matrix created by a convolution of the elements of current system parameters and the elements of innovation.

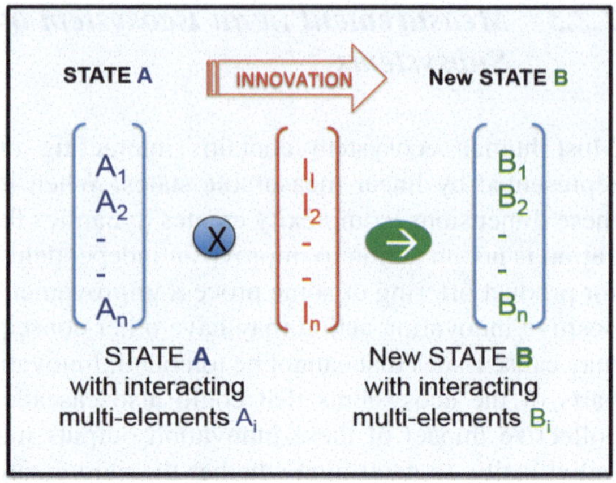

Fig. 2.8 Innovation affects many system elements in a nonlinear way

Measurements of the overall performance of the ecosystem may lead to information that shows unintended and unacceptable consequences for a certain segment of the ecosystem. This, of course, will require altering the initial innovation driver. Thus, measuring any of the innovation actions must be dynamic and iterative. Furthermore, measurement cannot be localized to the intended segment or subsystem. Analysis must be dynamic and be performed in a 'zoom in' and 'zoom out' modes. By zooming in, detailed information on the subject matter become available. By zooming out and casting a broad net that goes beyond the target, unintended consequences might be revealed giving information or direction for an important impact, both positive and negative.

In this context, big data methodologies might bring out some trends in a direct or indirect way (see for example Reshef et al 2011). Correlations and associations of data that are not initially well understood may lead to pattern recognition and bring out new unobserved trends (Fig. 2.9).

A single action of innovation, creating an output in a part of the ecosystem may also propagate a set of innovation actions in other parts of the ecosystem, each having its own dynamic. Some of these cascading effects can be unpredictable and may lead to dramatic social effects. A well-known example is the use of innovative social media on the Internet, intended mainly for social connections, but is also used to propagate political information and news leading to unrests and unearthing corruptions.

Another simple example illustrating an action propagating through the ecosystem is given in Fig. 2.10. A Government action propagates through the rest of the ecosystem: Academia, Industry and Society. In each subsystem an action is generated as a result of the Government trigger, and thus different innovations emerged in the ecosystem.

In addition to measuring quantifiable drivers and innovation actions there are classes of outcomes (could also be issues and challenges) that are not simply quantifiable with data. Experienced innovation leaders may 'sense' them and

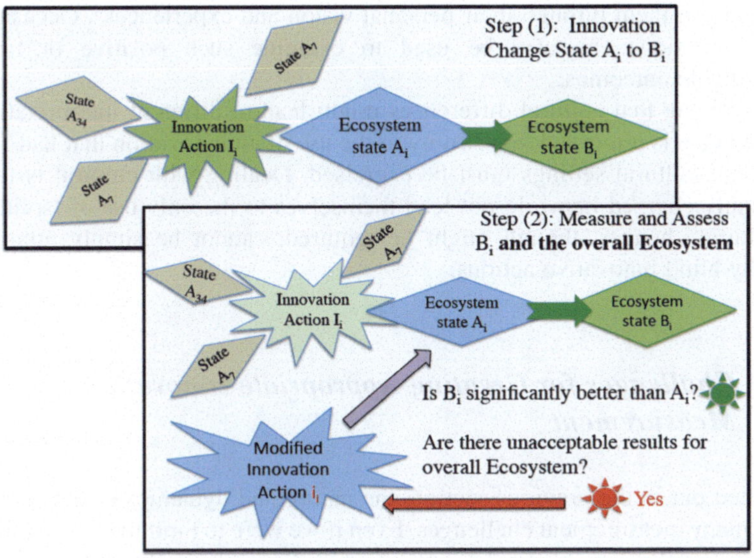

Fig. 2.9 An innovation action I_i intended to act on ecosystem element A_i, may affect other parts of the ecosystem such as A_1, A_7 and A_{34}. Assessment of action I_i requires a broader evaluation than B_i. If the outcome of the overall system performance is not acceptable, action item I_i must be modified

Government trigger The government seeks a reduction in CO_2 emissions.	**Industry** Industry is incentivized to generate a range of innovations around fuel efficiency. One industry player develops a more fuel efficient gas turbine.
Society Society respond by considering fuel efficiency as a priority purchase consideration.	**Academia** Academia collaborate with industry to advance gas turbine design – for example by whole engine modeling. They also adjust their curriculum to produce more graduates in relevant areas.

Fig. 2.10 An illustration of interacting parts of the ecosystem. A government action causing several different actions in parts of the ecosystems

make sense of them through their personal vision and experiences[2]. Occasionally, group discussions can also be used to examine such positive or negative unquantifiable outcomes.

Recognizing that cultural differences might lead to different interpretations of measured data is important. Sensitivity to the use of an innovation that had worked in different cultural settings must be exercised. Dealing with cultural issues and emotionally charged items do not lend themselves to the only use of 'solid data,' and cultural changes, though might be required, cannot be simply attained by culturally blind innovative actions.

2.2.4 *Challenges for Creating Appropriate Innovation Measurement*

As pointed out in the previous section, the nature and dynamics of the ecosystem present many measurement challenges. Even if we were to limit the measurement to few indicators that can inform the innovation leader on the health of the ecosystem, challenges are significant. Below we introduce few of these challenges for the purpose of making them visible and keeping them in check. In the subsequent sections we look at how these challenges can be addressed.

Challenge 1: Size and Scale of the Innovation Ecosystem The representation of the ecosystem highlights the innovation challenge: How can we measure the health of the innovation ecosystem when it has so many parts playing out in so many contexts? These contexts span geographies, cultures, problem spaces and different types of organizations from small nimble start-ups to old staid and large established organizations. In addition, there are several key entities within the innovation ecosystem such as government, industry, academia and society. Furthermore, the ecosystem allows teams to form around concepts and ideas, mobilizing resources through formal or informal projects that create and deliver value with a positive impact for people. All of these points present major challenges to the measurement system.

Challenge 2: The Dynamic Dimension to the Innovation Ecosystem How can we measure the evolution of the ecosystem? For example, how do measure change and rate of change leading to growth and sustainability indicators? As important as it is to know how healthy the ecosystem is today, we need to know how the system is evolving. Are parts of the ecosystem growing or shrinking? With measurement leading to making adjustments, it is, therefore, important to know the dynamics that are occurring. We also discussed the nonlinearity and interactions among the components of the ecosystem and showed that when and where innovation will occur is not predictable. Similarly, it is not predictable where innovation capabilities will decline.

---

[2] An analogy of that is a car driver who can tell the condition of the road without resorting to analytical data.

Challenge 3: The Ecosystem is an Integration of Several Parts and the Connections It might be possible to decompose the innovation ecosystem into parts that can be measured. But, as we discussed earlier, innovation propagates and thus there is a significant interfaces and interactions among these parts. So, will measuring parts give us a measure of health of the overall system? By way of a metaphor, if we were to deconstruct the human body into its parts and measure each part, would that tell us that the human body was healthy? In addition to measuring parts in the ecosystem, we need to be able to assess how the parts are combining, that is how the "white space" and connections between entities in the ecosystem is participating in creating value. In addition, how do we measure the emergent properties of the ecosystem that have no evidence when only examining the parts?

Challenge 4: Building on Existing Measurement Research There is significant existing measurement research on innovation in many areas of the ecosystem. How can we take advantage of this research and utilize it to enhance knowledge? Would a network approach work for most cases? Can one create a community of people who are interested in measurement to share experiences and tools? Which measurement areas are well developed and what other parts the approach to measurement is quite embryonic? How can we locate what had been measured?

Challenge 5: The Vital Signs of Ecosystem Health With the complexity of the ecosystem, can we find "vital signs" and we concentrate on measuring those, rather than measuring many thing? What are the "vital signs" that indicate if the ecosystem is in good health, and not in bad health? To take the human body metaphor a step further, there are thousands of tests and measures that can be undertaken to determine health. But there are a small number of critical measures such as breathing, a pulse and a lack of severed arteries and veins that will determine whether the human body ecosystem is going to survive the next ten minutes. And there another set of vital signs such as weight, blood pressure, eating habits and exercise habits that will inform health of the human body ecosystem over the next ten years. Therefore in measuring the innovation ecosystem it is essential to know what the vital signs are to avoid over measuring.

Challenge 6: Generality and Specificity We have a general model. How can we deepen the model by tailoring the generic concept for industry, academia, government and society? To be a useful framework, some translation is helpful into different contexts. Too generic will not be helpful but equally too specific will tie down the innovation, therefore destroying new ideas and opportunities.

2.2.5 Measurement of the Elements of the Innovation Cycle

As we discussed in section 2.1.4, innovation actions create knowledge and learning leading to innovation capability throughout the ecosystem. We decomposed the innovation cycle into several interacting elements: Innovation Activities leading to

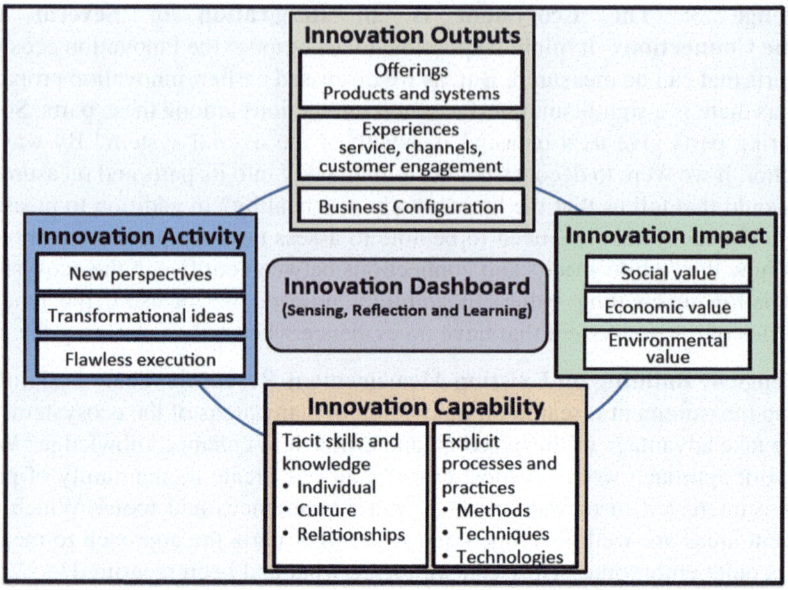

Fig. 2.11 Innovation Ecosystem is deconstructed into four components: Innovation Activity, Innovation Output, Innovation Impact, and Innovation Capability

Innovation Capability, Innovation Output[3], and Innovation Impact. Each component has rich elements and deliverables and contributes to the creation of innovation knowledge that enriches the ecosystem and leads to innovation capability. Such capability might be decomposed into human, processes, infrastructure and culture. One can ascertain that the innovation capability is the most important output of the innovation cycle and resides at equal footing with the outcomes. Ecosystems that develop capabilities in a systematic and consistent fashion end up at the top of the food chain. In fact, one hopes that innovation leaders pay significant attention to measuring the capability of the system and monitor innovation directions that enhance the growth of the innovation capability (Fig. 2.11).

An advantage of this model is that it focuses the emphasis of the elements of the ecosystem into a theme. Yet each of these themes is diverse in scope and outreach, and thus different schemes and tools need to be considered in measuring their effectiveness and ability to contribute to success in reaching the desired state as well as increasing the innovation capability. And although there is a risk in assuming that the innovation ecosystem can be deconstructed into independent components, the model is a reasonable first approximation for a division that allows independent measurement, and gives specificity to the tasks in the innovation cycle.

As we discussed, the model of the innovation cycle consists of four elements:

[3] The concept of innovation output was discussed by (Keeley et al. 2013).

- The *innovation activity* might produce new perspectives, transformative ideas and methods, and processes to perfect the execution of all aspects of the innovation cycle. Measurement of an activity can be related to a particular standards and norm, and it might be possible to quantify and scale the activities and might be possible to do some benchmarking.
- The *innovation outputs* contain products and systems that might be tangibles; some of the human experiences connected to these tangibles and some are not. Then there is the innovation in business configurations with its richness and ever evolving nature. Here measurement, for example, can be devices to relate to some engineering and economic parameters as well as customer satisfaction.
- *Innovation Impact* is not easy to measure. Some impact might be fast and long lasting. Others can manifest success is more futuristic. Then there is the impact that can be elusive and show up in an indirect way and in some unexpected domains. The domain that impact can influence also varies from social, economic and ecological including sustainability, resilience and related issues. Here we can easily connect to complexity and 'wicked problems' with their large-scale human issues. Another point worth mentioning is that impact might not be localized in a certain domain and it might be a root cause for some future successes that may take some time to surface out in some domains.
- The *innovation capability* of the ecosystem is the depository of the innovative experiences of the ecosystem. Measurement of the evolution of the system capability is critical.

It should be noted that in the previous discussion the ecosystem was presented with minimum complexity, and thus one may perform measurement as if the elements of the ecosystem were independent, but we know this is not correct. One cannot ignore the feedback and feed-forward loop and innovation leaders must be very alert not to optimize along local successes.

An illustration of some 'innovation loops' can be observed from the following diagram presented in Fig. 2.12.

- The left side of Fig. 2.12, the loop presents that the 'experience' of an activity enriches the capability. This causes new activities to be generated from

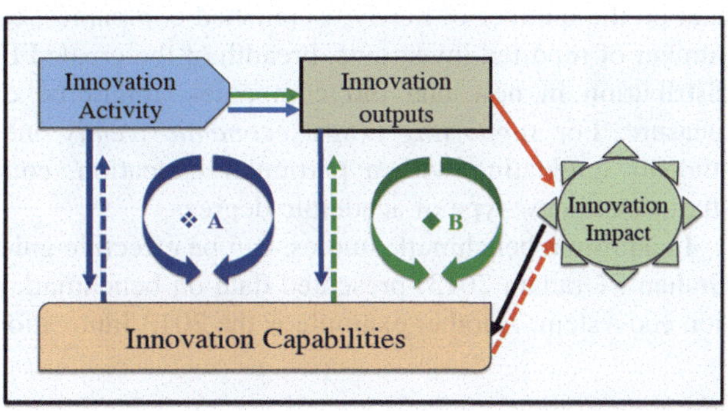

Fig. 2.12 An illustration of innovation loops in the innovation cycle model

accumulated capabilities. All of these activities convolute to create output and impact.
- The right side of Fig. 2.12, the loop presents: (i) An activity leads to an output that creates an impact and increases capability. (ii) An output(s) add to another output(s) creating a new capability that leads to a new output and new impact.

When measuring the parts in the innovation ecosystem, as described above, the following should be considered:

- The measurement should bring out indicators of positive health as well as indicators of negative health. For example, large organizations may have a culture that actively fights back at suggestions for improvement, therefore reducing the flow of new perspectives and transformational ideas. Strong project management and execution skills may exist, but without new perspectives and transformational ideas the projects are focused in the wrong areas or may result in marginal improvements.
- Indicators should involve a point of comparison to determine what the indicator means. This point of comparison could link to an international benchmark, a previous measurement to give trend over time or an absolute number or percentage.

2.2.6 *Vital and High Priority Indicators*

Monitoring the health of innovation and measuring its progress in reaching a desired state is critical. A particular innovation might be on a particular path to reach a desired state; obstacles or changes in some conditions of the ecosystem might divert its course. Guidance and course corrections are needed. There are vital signs that provide an overview of the impact of a particular innovation and the performance of the ecosystem. However, there are no universal vital signs that one can use, and it is important to opt measuring parameters and tools that are suitable for the particular segment understudy (Fig. 2.13).

For example, if the area of interest is the *creation of new economy*, obtaining data on the number of newly established companies, earnings from new products, number of reported inventions, breadth of the created IPs, and employment and its distribution in new and old companies, might be examples of parameters to measure. For *supporting future economic trends* one may measure number of students graduating with a particular education, courses with interdisciplinary studies, and new type of academic degrees.

In addition, benchmark studies, can be effective guides for obtaining data. Ruth Graham (Graham 2013) presented data on benchmark study for technical innovation ecosystem. Another example is the 2013 Innovation Barometer[4] conducted by

[4] Check: https://www.ge.com/sites/default/files/Innovation_Overview.pdf.

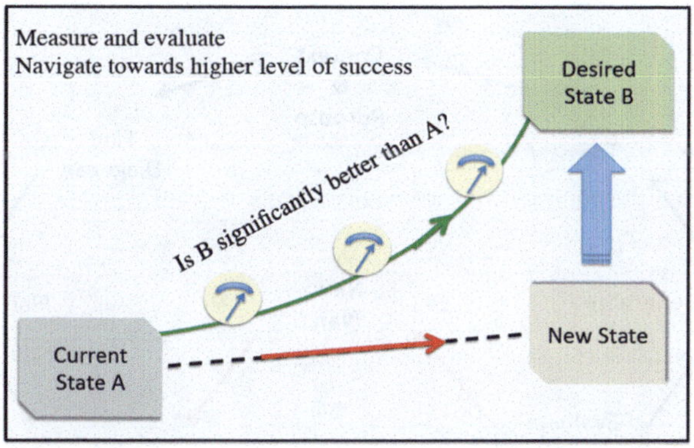

Fig. 2.13 Measurement as a tool guiding the innovation ecosystem to reach desired goals

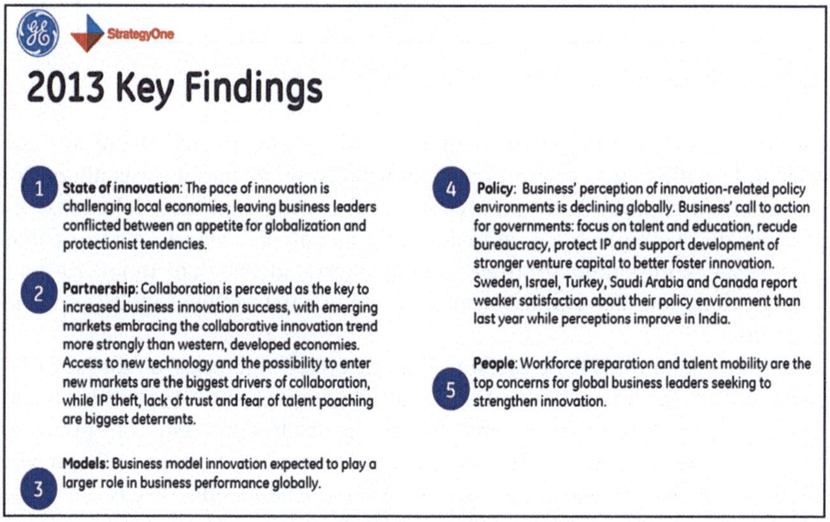

Fig. 2.14 Key findings of the GE Innovation Barometer study

GE. In this survey 3,000 business executives in 25 countries gave their opinions on drivers and deterrents to innovation.

Out of this survey, important innovation indicators surfaced:

- Understanding customer latent need
- Attracting and retaining innovative people
- Ability to develop new technologies and processes
- Create innovation environments and cultures
- Working collaboratively with business partners through the ecosystem

Their key findings are listed in Fig. 2.14. An important aspect that was mentioned is the speed of innovation and its effect on the ecosystem. And although

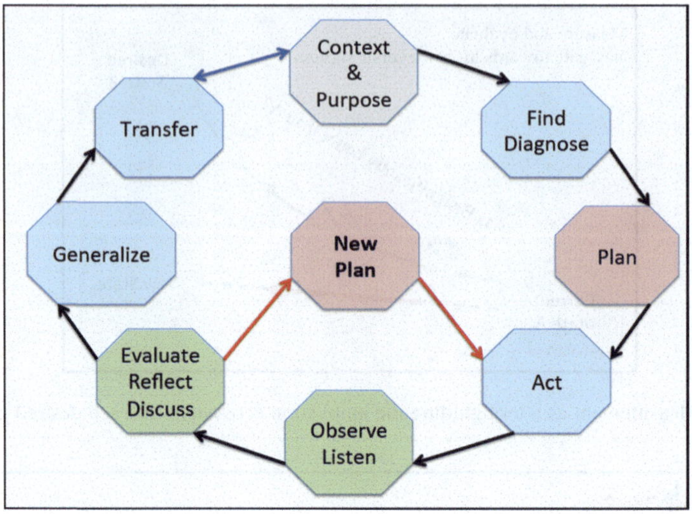

Fig. 2.15 Action Research process

rapid development and deployment of innovations are favored, there are consequences and conflicts that require management. Another finding was about policy and government support to the innovation ecosystem. Regions might loose competitiveness due lack of policy and economic incentives. Measurement of that on regional basis is important. Further finding is related to talent mobilization and talent migration, and the need to monitored and understood the root cause of talent demography.

In addition, other tools for understanding the effect of innovation (or lack of) can be gleamed from the human system. After all, the impact of innovation is on people and society. Providing environments for group discussions and communications might not provide quantitative data, but insights of what is working well and what is not. Articles by Kemp and Pearson, and Kesidou and Pelin discuss drivers for innovation and give insights for measuring them.

'Action Research'[5] used in education can be a method for reflection and analysis. Whereas academic research aims at finding general solution, action research gains knowledge that can be directly applied to a particular situation. The process starts with identifying a problem, devising a plan and implementing it, then examining and reflecting on the entire process. Another cycle starts with the opportunity for continued reflection. Action Research provides innovative leaders with a process to examine their own work and performance (Fig. 2.15).

[5] See for example articles in the "Action Research Journal": http://arj.sagepub.com, and http://infed.org/mobi/kurt-lewin-groups-experiential-learning-and-action-research/

2.2.7 *The Measurement Dashboard: A Strategic Navigation Tool*

As discussed in section (2.1), an innovation activity draws on a knowledge base to generate outputs, and through a self-generating cycle the innovation capability grows. This innovation capability resides within the entities of the ecosystem such as government, industry, academia and society.

In the center of the innovation generation cycle is the measurement dashboard providing information for the leadership to evaluate and monitor the health of the innovation ecosystem as well as to navigate innovation actions towards successful outputs.

Measurement dashboards are not easy to construct, but they are essential for understanding the status of the ecosystem. Innovation leaders need to help *establish* such measurement tool in the different elements of ecosystem. Dashboards provide innovation indicators to determine not only the health of an entity of the ecosystem, but also the magnitude of the transmitted innovation experiences and learning among the entities. Thus building the ecosystem innovation capability becomes a collective action among the innovation entities. Russell et al (2011) bring important concepts on methods for transforming innovation ecosystems through shared vision and networking which need to be exercised thought out the entities of the ecosystem (Fig. 2.16).

The resulting innovation capability, created in a given entity is diffused and transmitted to other entities and an overall ‘innovation intelligence’ emerges. Monitoring and driving this process of creating, growing and sharing is one of the most important contributions of the innovation leader. By creating forums and

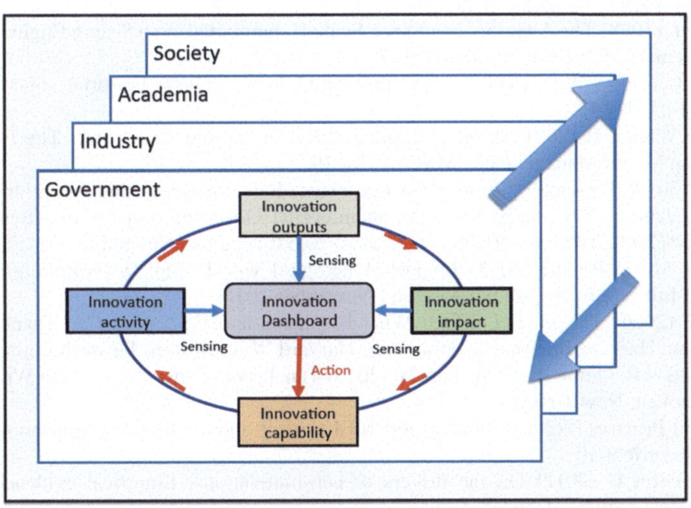

Fig. 2.16 Innovation capabilities are to be transmitted across the ecosystem entities

methods for exchanging and merging innovation capabilities across the ecosystem, the overall ecosystem creates an upward moving spiral.

2.2.8 Conclusions

This section emphasizes the importance of creating set of measurement and innovation health indicators as well as taking advantage of tools already tested. It also attempted to bring out the interacting elements that create complexity. In addition, the importance of creating measurement dashboard to help innovative leaders navigating the ecosystem to successful outcome is outlined. Special care was recommended to integrate the outputs of the dashboards of different ecosystem entities leading to collective creation of innovation capability. Most important is to realize that although developing learning and conclusions from quantitative and qualitative measurements is important, innovative leaders must be able to 'sense' the condition of innovation in their ecosystem and be able to make decisions based on measurement as well as inner sensing. This sensing is an internal know-how and a tacit knowledge developed by experience and by being attentive, thoughtful and mindful about the overall ecosystem.

References

Berkeley College of Music, 'Rethink Music. New business model in the music industry,' A framing paper for the 'Rethink Music Business Models Workshop.' April, 2013. Valencia, Palau de les Arts, Valencia, Spain.

Brin S, Page L (1998) The Anatomy of a Large-Scale Hypertextual Web Search Engine. Computer Networks and ISDN Systems 30:107–117

Churchman C (1967) West used "wicked problems" in his "Guest Editorial" of Management Science 14:141–142

Fox M, and Wrenn, B (2001) A broadcasting model for the music industry. The International Journal on Media Management, JMM 3:112–119

Graham R. (2013) Technology Innovation Ecosystem Benchmarking Study: Key findings from Phase 1', January 2013. http://www.rhgraham.org/RHG/Recent_projects_files/Benchamrking%20study%20-%20Phase%201%20summary%20.pdf accessed November 2014.

IFPI Digital Music Report (2013) Engine of a digital world. http://www.ifpi.org/downloads/dmr2013-full-report_english.pdf accessed November 2014.

Johnasson F (2006) The Medici Effect: What Elephants and Epidemics Can Teach Us About Innovation. Harvard Business School Press. Harvard Way, Boston, Massachusetts.

Keeley L, Pikkel R, Quinn B and Walters H. (2013) Ten Types of Innovation. John Wiley & Sons, Inc., Hoboken, New Jersey.

Kemp R, and Pearson P (2007) Final report MEI project about measuring eco-innovation. UM Merit, Maastricht 10.

Kesidou E, Pelin D (2012) On the drivers of eco-innovations: Empirical evidence from the UK. Research Policy 41:862–870

Leonard DA., and Straus S (1997) Putting Your Company's Whole Brain to Work. *Harvard Business Review* 75, no. 4 (July–August 1997): 110–122

Leurdijk A, de Munck S, van den Broek T, van der Plas A, Manshanden W and Rietveld E (2012) Statistical, Ecosystems and Competitiveness Analysis of the Media and Content Industries. Jean Paul Simon ed., European Commission, Joint Research Centre; Institute for Prospective Technological Studies. http://ftp.jrc.es/EURdoc/JRC69435.pdf

PricewaterhouseCoopers 'Global Entertainment and Media Outlook' in http://www.pwc.com/gx/en/global-entertainment-media-outlook/, accessed in November 2014.

Scholes R (2000) Uniquely GoogleTM, Stanford Technology e-Brainstorm. The Newsletter of Stanford's Office of Technology Licensing (OTL).

Reshef DN et. al. Detecting Novel Associations in Large Data Sets, Science 334, 1518 (2011).

Rittel H, Webber M (1973) Dilemmas in a general theory of planning. Policy Sciences 4:155–169

Russell MG, Still K, Huhtamäki J, Yu C and Rubens N (2011) Transforming Innovation Ecosystems Through Shared Vision and Network Orchestration' in Triple Helix 9 – Innovation Ecosystems Summit 2011 – Stanford University.

Vise D and Malseed M (2005) Google Story Inside the Hottest Business, Media and Technology Success of Our Time in 'The Google Story', Bantam Dell Publishing, NY.

Chapter 3
Innovation Leadership: A New Kind of Leadership

Banny Banerjee, Stefano Ceri, and Chiara Leonardi

3.1 Introduction

The previous chapters on innovation and the innovation ecosystem lead us to the central topic of this chapter: Innovation Leadership. Despite the many variables influencing creativity and innovation in organizational settings, leaders and their behavior represent a particularly powerful influence (Mumford et al. 2002; Sarros et al. 2008). Traditionally, leadership has implied a position of power, carrying out executive functions and delegation of tasks to subordinates in hierarchical organizational structures. Currently, leaders emphasize risk-averse decision making and efficient management of resources, resulting in incremental changes rather than radical pathbreaking innovations.

As we have seen in Chap. 1, innovation is not simply the creation of a novel concept or an extemporaneous reaction to a situation (Banerjee 2009). It has a distinctive theoretical basis, a structured process and a unique cognitive style and epistemology (Cross 2006), which give it the power to transform systems and their trajectories. The next generation of leadership will not only be defined by characteristics that are importantly different from current forms, but will be framed by an entirely different paradigm that shifts the focus from the "individual as leader" to individuals occupying "leadership modalities". In order to solve complex challenges innovatively, we need a "step change" in the concept of leadership, moving far beyond the traditional references to military-like hierarchy, or centralized business management towards a more distributed model of leadership that

B. Banerjee
Stanford ChangeLabs, Cordura Hall, Rm. 104, 210 Panama St., Stanford, CA, 94305-4115
USA

S. Ceri (✉) • C. Leonardi
DEIB, Politecnico di Milano, P.za L. Da Vinci 32, 20133 Milano, Italia
e-mail: stefano.ceri@polimi.it

B. Banerjee, S. Ceri (eds.), *Creating Innovation Leaders*,
Understanding Innovation, DOI 10.1007/978-3-319-20520-5_3

unlocks the capacity for an organization or a community to innovate in entirely new ways.

Innovation Leaders have the vision, competencies, and skills that would not only make themselves effective, but also be important in scaling innovation capacities. They have agency and are in a position to influence the vision, the strategic direction, and the value systems, practices, decisions, and actions. Innovation Leaders are responsible for shaping alternative futures rather than simply managing the present problems with normative methods. Hence Innovation Leadership needs to emerge as a discipline that embodies heterogeneous teams (Somech 2006), trans-disciplinarity, creating new cultures, and catalyzing synergy across institutions and organizations in ways that generate new possibilities, forging pathways out of decision gridlocks. Both deep and broad knowledge are implied in a leadership that nurtures plurality of approaches, modes of inquiry, and conceptual underpinnings. It falls on the Innovation Leader to draw on a broad diversity of disciplines, theories of success, tools, and techniques in order to build bridges between current challenges and desirable futures.

3.2 Next Generation Leadership

The term leader has been often characterized by an attitude of a commander, imposing ones vision and decision making power upon subordinates, which propagates hierarchically in organizations. (Stogdill et al. 1957) In contrast, today's leaders must not only be brilliant strategists, creative decision makers, and effective motivators regardless of the field they are engaged in; they must be also able to imagine new scenarios and creatively conceive solutions that go beyond established ideas and norms. They must identify the right questions. Moreover, they must be able to achieve their leadership on the field, through an ability to interact with project teams and by successfully promoting their ideas in the context of open and cooperative processes which go beyond hierarchical lines of command. Furthermore, innovative leaders must be sensitive to the complex social and environmental issues that now represent the grand challenges, and develop the means of co-creating solutions with members from different disciplinary or agency domains.

If we look at some of the most remarkably successful leaders across sectors such as Steve Jobs, Momammad Yunus, or Martin Luther King, their behavior often does not map with traditional models of what is considered leadership in the context of an organization or institution. Transformational leaders inspire, energize, and intellectually stimulate others (Bass 1990).

Figure 3.1 is a depiction of Innovation Leadership as a simultaneous increase in the ability to address more and more complex challenges that are marked by scale, urgency, and importance along with the ability to influence deeper and broader impact. We will discuss the nuances and implications of this shift later in the chapter, but this framework is an important lens with which to view leadership in the future because it casts leadership in its ability to transform, rather than in terms of a position in a hierarchy. We believe that the emphasis on impact and

Fig. 3.1 Innovation Leadership catalyzes and enables creative approaches increasingly scaled and complex challenges with deeper and more scaled impact

transformation is a key pivot in the way leadership is perceived and a fundamental driver of all thought and action carried out downstream of having the shift in mental model.

If the role of the leader is not just to excel in the current paradigm, but also to replace the current paradigm with more appropriate ones, it has some deep implications. It implies that the leader has a responsibility to *disrupt* current models and paradigms with more appropriate ones. This in turn implies an ability to escape the current paradigms and to question conventional wisdom amidst power structures that are built around conventional wisdom. This can only be done with a great deal of original critical thinking and systems thinking. If a new kind of thinking is a dimension along which the capacity is expanded, then the other side of the coin is *a new kind of action.* Current processes and implementation methodologies are built on top of deeply entrenched mental models and beliefs of the conventional ways of thinking, and consequently carry with them the *theoretical limits* that a new kind of action would have to transcend.

3.2.1 Innovation Leadership as a Modality

The next generation of leadership is not going to be simply embodied in an organizational position or necessarily be a function of seniority, but a modality that involves fulfilling certain functions in the context of an organization, institution or project, as shown in Fig. 3.2.

Innovation Leadership is not simply an ability to manage, administer, and make executive decisions, but involves an enhanced emphasis along a number of other dimensions listed below:

Fig. 3.2 Innovation Leadership is a modality that any individual can occupy by exhibiting the various capabilities that are emblematic of this emerging type of leadership

- **Envisioning alternate futures:** The ability to think in original and creative ways about desirable futures, both in the short term and the long term, that one can then direct one's action towards.
- **Reframing intentionality and success**: The ability to escape conventional wisdom and redefine the goals, the objectives and the thinking around what constitutes success.
- **Creative judgment**: The ability to assess and judge in ways that are sensitive to multiple perspectives, so as to deal with complex multi-dimensional challenges with nuances and subtle implications.
- **Strategic Decisions:** The ability to drive processes towards highly strategic decisions that are sensitive to human behavior, social systems, economic or business systems, environmental sustainability, and resilience of systems in the face of changes.
- **Inspiring action:** The ability to create a bias for action as a basis for thought, alongside thought being the basis for action.
- **Organizational transformation:** The ability to change the culture, values, practices, position, purpose, and efficacy of an organization.
- **Transforming an ecosystem:** The ability to have a secondary influence on the entire ecosystem of stakeholders and interrelated outcomes rather than on just one's own organization.

Adam Lowry and Eric Ryan From Method

Revolutionizing the home cleaning products industry through human-centered innovation

The home cleaning products sector is a multi-billion dollar industry. Many of its central ingredients were adapted from chemicals developed during the World War II and repurposed for other applications. Thus, conventional cleaning products have a high toxicity and negative environmental impact. Laundry detergents in particular are significant contributors to a deteriorating stream of profit and the coastal ocean ecosystem health. But, with a reliable consumer base and comfortable profit margins, "innovation" wasn't a common word in the industry and there was very low impetus for change.

When former college roommates Eric Ryan and Adam Lowry looked around the sector, they saw that a few "green" cleaning products already existed. But it was a tiny percentage of the market the catered to a niche of self-identified environmentalist consumers, and aesthetics wasn't their priority. "Back in 2001, green cleaners were just hideous looking", the 38-year-old Ryan says. And the conventional wisdom was that green didn't clean. "Simply using more eco-friendly ingredients and labeling it as 'green' wasn't going to change consumer behavior in any significant way." Ryan and Lowry posed themselves a challenge: create an eco-friendly cleaning product that a wide diversity of customers would actually desire. They dove into product development with the intuitive insight that people's homes are deeply tied to a sense of personal identity and aesthetics.

Could they create a cleaning product the reflected that? Lowry, with a degree in chemical engineering and a background in environmental science, formulated a dish soap and a line of all-purpose cleaners that were nontoxic and biodegradable. The sprays had pleasant fragrances like cucumber, a rarity at that time. "To be successful, we needed to bring the mainstream into green cleaners. The big idea was to blur the lines between personal care and home care. That's where a lot of the design, the fragrance we use, the personality comes in. We wanted to bring fun into it. At the end of the day, the environment doesn't care what your motive for buying it was", explained co-founder Eric Ryan. "Method is now the fastest growing eco-friendly home products company in the world".

(continued)

The team prioritized sleek, minimalist design and soothing color palettes, drawing heavily from principles used in personal cosmetic products. Through talking and testing with consumers in Method's early days, Ryan and Lowry realized that people weren't choosing their products because of the ingredients, they were choosing them because they matched the color of the soap with the color of their bathroom walls, or the scent reminded them of the relaxing luxury of a spa. A partnership with the designer Karim Rashid led to Method's big break, a 2002 distribution deal with Target. Next year, Method debuted a hand soap in what would become the brand's distinctive teardrop-shaped bottle which earned Method their first break into the home products market. Ryan and Lowry demonstrated leadership throughout their work, from setting up a small initiative to driving it to worldwide success, by changing the paradigm of the cleaning product and by being able to attract collaborators and build partnerships.

Framing Innovation Leadership as a modality implies a radical frame shift that implies that Innovation Leadership is not just exercised by people of political or organizational seniority, but could be exhibited by anyone in any context. Given that Innovation Leadership is a modality, it can be assumed by different people at different times in the same team or organization in such a way that everyone can be expected to be a leader, instead of the role of a leader being associated with an individual. This model opens up the door to a distributed model of leadership where multiple agents assume the leadership modality, and the sum of their efforts could be far greater than what could be achieved by one individual.

In a hierarchical organization, much of innovation capacity remains under-leveraged because individuals lower in the ranks do not have the agency to exercise it. They are expected to follow rules, conform to the decisions coming in from the top, and toe the line, thus undermining a culture of co-creation (King and Anderson 1990). Hierarchical leadership is not only wasteful in leadership capacity but it is also inefficient. Each step for leadership command carries the potential for *"drift"*. When the concept of *leadership* is dissociated with *organizational seniority,* then everyone has the license to exercise the leadership modality and the organization or community is far better geared for focusing on solving challenges of increasing complexity (Banerjee 2014).

An approach known as Complexity Leadership Theory argues that three types of leadership can be observed in organizations: administrative, adaptive and enabling. The third perspective builds upon the evolutionary theory of the firm (Uhl-Bien et al. 2007) and looks at leadership as an organizational meta-capability that manages other capabilities that in turn are engaged in both exploration and exploitation. In particular, Complex Systems Leadership Theory defines leadership as a system function that operates to "changes the rules of interaction" among both people or groups within a complex adaptive system of interactions, in terms both of ends - where the system is going - and means - how to get there (Hazy 2007; Goldstein et al, 2010).

Leila Janah and SamaSource

Innovation Leadership adopting innovation as a new modality

We live in a world with vastly unequal distribution of opportunity, with 1.4 billion people living on less than $1.25 a day and several more billion people living on less than $2 to $3. Development programs aimed at poverty eradication find it difficult to create employment where the economies do not support any livelihood or employment opportunities.

Leila Janah, 29, is a social entrepreneur who uses technology and lean business methods to promote social and economic justice in a new innovative way. She got inspired by her experiences with the World Bank and in conducting fieldwork in Mozambique, Senegal, and Rwanda. In 2008 she launched Samasource, a non-profit social business that gives digital work to impoverished people around the world.

Rather than using traditional development models, Leila Janah is harnessing the power of the Internet to change lives across the world: she realized that one way to impact sourcing is to take outsourced digital work from developed economies and give it to people living in poverty in under-developed regions around the world. Her innovative model uses the internet to bridge the gap between people in desperate need of livelihoods, and to established markets in the west. Samasource finds companies struggling with online tasks such as comment moderation, transcription, and tagging and connects them with workers living in places with extremely high unemployment and trains them to carry out these digital tasks. Since the most complex projects are composed of small tasks, they can be simplified, distributed and performed by workers who can be trained in basic computer skills through a technology platform.

In this way Samasource provides income and educational opportunities for marginalized workers in slums, refugee camps, and impoverished communities across Africa and Asia, moving more than 20,000 people over the poverty line.

Leila Janah's company, Samasource, is changing lives of unemployed people in India, Uganda, Kenya, Ghana and Haiti

(continued)

Janah, thanks to her great innovation leadership, was named one of the Most Influential Women in Technology by Fast Company in 2009 and received a World Technology Award for Social Entrepreneurship. Innovation Leadership here is not just exercised by people with organizational seniority. Leila demonstrated Innovation Leadership by creating a highly innovative paradigm to address the intractable challenge of people caught in poverty traps in regions without economic opportunities.

3.2.2 Innovation Leadership as Maximizing Innovation Capacity

At the most abstract level, the job of the Innovation Leader is to amplify impact with the greatest sphere of influence possible, and raise the innovation capacity within the system, whether it is in the context of a project, a team, an organization, an institution or a large transformative initiative. In order to fulfill their fundamental purpose (which is to create a continued capability of bringing about appropriate transformations with disproportionate efficacy), Innovation Leaders have to worry about two interrelated but separate issues: The first question for an Innovation Leader to pose is "How to bring about amplified impact", quickly followed by the question "How to create the continued capacity to create amplified impact". The fundamental pathways through which an Innovation Leader carries out this function thus fall in these two categories:

1. **Amplify the impact**
   (a) By crafting transformative visions, proposing new directions, reframing success, creating a space for transformational discourse, crafting new paradigms, and developing new conceptual bridges.
   (b) By creating new ways for actualizing concepts and driving impactful transformations.
   (c) By amplifying the scale, nature, depth, and the pace of system-wide transformation and the nature of impact.
   (d) By influencing what types of challenges are targeted, setting the direction for actions, driving new outcomes, generating new theories, models, processes, platforms for advancement, and resources to drive impact, thus having an impact beyond this direct scope of influence.
2. **Amplify the** *innovation capacity* **of the system**
   (a) By catalyzing a culture of innovation and building an ecosystem that fosters innovation.
   (b) By diversifying the types of innovation to more unbounded challenges (moving from Class A challenges to Class C Challenges as introduced in Chap. 1).
   (c) By creating the scaffolding for innovators to gain expertise and raise the average level of innovation expertise across the ecosystem.

3.2.3 Innovation Leadership as a Mindset

A mindset is the larger frame of assumptions, entrenched methods, decision and motivational stances; in general systems theory; it also refers to the larger set of background assumptions by a group of people who influence choice or behavior. Thus, seeing Innovation Leadership as a mindset has many implications along with the purpose of making continued impact, which is the central purpose of an Innovation Leader.

The term "Perceptual Lens" was originally proposed by Egon Brunswik (Brunswik 1952) and later developed by Kenneth Hammond (Hammond 1980), among others. It is a profound construct that is an implicit component of any field and is a critical factor to consider, especially in a context in which there are multiple motivations and epistemologies at play.

It is important to understand that any discipline or organization develops a certain bias for what it cares about and what it doesn't. Along with the bias for what it cares about, it develops a selective ability to see certain things and be oblivious to others. This bias of being perceptive of some things and not to others creates a "Preceptual Lens" and has a deep impact on process, decision making, goals, and definitions of success.

Imagine a scene in a movie where there is a tense moment depicting an altercation between the captain of a submarine and a subordinate about a critical decision in the midst of mechanical malfunctions and broken communications. Even while being immersed in the movie, a mechanical engineer will speculate about hull integrity, the communications person will note the signal to noise ratio, the cognitive psychologist will recognize confirmation bias in the captain, the fashion designer might note the authenticity of the World War II uniforms, and a historian might be aware of the historical context in which this battle is being waged. It is unlikely that the cognitive psychologist will worry about stresses in the hull, or the mechanical engineer about the psychological nuances. We notice what we are looking for, and what our perceptions are honed to observe at the cost of other details. To a person who does not care about color, the sky in a given moment might be blue. To a painter, it might be powder blue, with a hint of cobalt blue and with tones of cyan.

One of the biggest issues arising out of members of a discipline trapped in their own Perceptual Lenses is that in a cross-disciplinary setting, since they are less perceptive to other people's points of view, there is a tendency to devalue the other person's opinions with an assessment that they cannot see the more important concerns. The Innovation Leader needs to internalize the concept of Perceptual Lens and be adept at working with multiple Perceptual Lenses in a trans-disciplinary context and create a space where the different perspectives can exist in a pluralistic, non-oppositional manner. Instead of different Perceptual Lenses being a problem to contend with, it is to be seen as an advantage, since the presence of different perspectives creates a "*parallax*" and allows a more multifaceted understanding of a complex situation.

While Innovation Leadership represents a unique mental model and approach, it can be overlaid onto any discipline or specialization. The ability to create a bridge between an open-ended question and an implementable solution requires a mindset that establishes a certain set of dualities: inquiry with action, sensitivity with gumption, analysis with synthesis, critical thinking with hands-on ability, theoretical thinking with hands-on exploration. The mindset is the fundamental operating system of the creative thinker; hence creating the right attitudes and philosophical stance becomes a very important function for an Innovation Leader.

The innovation mindset requires a willingness to "see" possibilities and imagine scenarios that are different from the current ones. It requires a drive for impact, while maintaining a comfort with ambiguity. Innovators play at the nexus of concerns, so they must understand the implications of research content, how technology might be leveraged, how to unpack unstated human needs, see ways in which business or market opportunities can be leveraged, understand the implications of broad socioeconomic backdrops, and know how to "make it happen". Above all, the Innovation Leader drives change from the status quo to a preferred state, and hence the fundamental role of an Innovation Leader is to be a change agent and a catalyst for change agency. This implies that an innovator's worldview, and ethical stance is as relevant as the capacity to innovate. The innovator leader's mindset is comprised of and should emphasize the following stances:

1. **Duo-temporal**: A view that places importance on both the short-term exigencies and the long-term resilience and sustainability of the solution.
2. **Systemic**: A perspective that embraces the complexity of interlaced systems and seeks points of intervention that are leveraged and will cause systemic transformation (Shipton et al. 2005).
3. **Holistic:** An approach that looks at multiple aspects, objectives and the diversity of points of view that comes with trans-disciplinarity.
4. **Growth Mindset:** A drive for continuous personal growth and innovation capacity of the team or organization, by actively internalizing and embodying a "growth mindset" (Dweck 2008).
5. **Non-determinism:** Recognizing that the agent of change is acting within the context of change (Findeli 2001) as a catalyst open to a variety of possible interpretations and adjusting with emergent developments.
6. **Dyadic thinking**: Top-down meeting bottom-up, strategic meeting ground realities, global needs meeting local conditions, practicalities meeting aspirational goals (Banerjee 2014).
7. **Exploration/openness:** Solutions to grand challenges require that we look beyond the currently available solutions, join dots in new ways, and explore avenues that might have the potential for radical change (Rosing et al. 2011). Often these challenges fall in the category of "unknown unknowns" where an explorative, iterative prototyping is a way of understanding what the *problem really is*. This approach requires a commitment to new possibilities challenging conventional wisdom, and pivoting around emerging insights (Valkenburg and Dorst 1998)
8. **Value Creation:** A deep commitment to taking on the innovation challenge of creating genuine value in the face of competing or multifaceted demands.

9. **Co-creative:** An ability to *lead from behind* and enable diverse teams to co-create solutions leveraging collective wisdom while averting a reductive and combative discourse.
10. **Insight-Based:** The humility to make the team realize that "*we don't know what we don't know*", with a belief that good decisions can only be based on deep insights, at the human and system levels.

3.3 The Duality of Innovation Expertise and Leadership

Like any other discipline, innovation has an expertise ladder that any innovator has to climb (Cross et al. 1994). Gaining expertise at innovation is unlike gaining expertise in fields where amassing content knowledge can be equated with being an expert. Gaining expertise in innovation is like becoming an expert at a sport like soccer. There is the theoretical knowledge of the sport that is very important, but no amount of theoretical knowledge supplants the ability to actually play a world-class game of soccer. Similarly, expertise as an innovator is not just gained through reading books about innovation, but through the experience of actually having done innovation work in the face of complex challenges.

At some level, many humans have the capacity to be innovative or to bring innovation to their discipline. For example, highly trained software engineers with a lot of experience in software programming might become facile with the task of creating software programs, but whether they are being innovative in their approach is another story. Two software engineers who have an equivalent level of experience and expertise might display very different levels of innovation.

There are people who bring innovation to a discipline that already has an identity, such as computer science or material science. And then there are others who work in fields that are tightly associated with innovation itself; practitioners in these fields are expected to perform on the dimension of innovation. Fields such as product design, interaction design, automobile design, fashion design, and architecture require that the practitioner be innovative and this expectation is a defining component of their identities and roles.

Expertise can grow along many dimensions in any discipline: the content knowledge, theoretical frames, technical proficiency, knowledge about processes, a culture of practice, the level of craft, communication protocols, taxonomy, language of the discipline, the ability to frame the problem and solve it, and the tools to realize conceptual solutions are all dimensions along which expertise grows for any practitioner. This happens as practitioners go through their career journeys and amass more experience, knowledge, skills, and strategic vantage.

With greater expertise comes the formation of higher-order heuristics, the ability to see larger patterns, the ability to judge or sense where the better solutions lie, and ultimately the ability not just to follow the existing practice but also to advance it. Expert chess players not only possess a vast knowledge of chess theory, history, styles, and combinations, but they also develop meta-cognition for the game

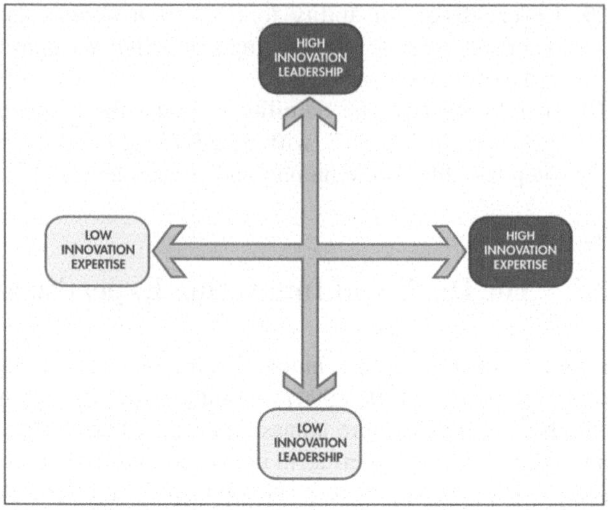

Fig. 3.3 A framework depicting Leadership levels versus Innovation Expertise

leading to deep intuition about the larger patterns of the game. The higher order intuition and meta-cognition that comes with expertise is particularly important in the context of problem types that do not lend themselves to deterministic frames and demand more interpretive approaches.

Although most Innovation Leaders would typically have a high degree of innovation expertise themselves, the two terms *Innovation Leadership* and *Innovation Expertise* are not to be confused with each other (Fig. 3.3). It is certainly possible for someone to be a tremendously effective Innovation Leader and raise the level of innovation in an organization or initiative without being an expert innovator and conversely, it is also possible for someone who has a very high degree of innovation expertise in a given type of challenge to be a very poor Innovation Leader. It is akin to being an Olympic coach of a gymnastics team without being a gymnast oneself while on the other hand, not every Olympic gold medalist in gymnastics would make a good coach.

Therefore, the notion of *Innovation Expertise* is an important construct for an Innovation Leader to grasp and ultimately influence. An Innovation Leader has to manage and grow a portfolio of innovation expertise across different specific domains or responsibilities.

3.3.1 The Innovation Matrix

Figure 3.4 is a framework that depicts the Innovation Matrix. In this diagram, the X-Axis represents the Challenge Class (as described in Chapter 1). Challenge Class A is "Implementation or Adaptation Type Problems"; Class B is "Open-Ended Medium Scale Design Challenges" and Class C is "Scaled Transformation

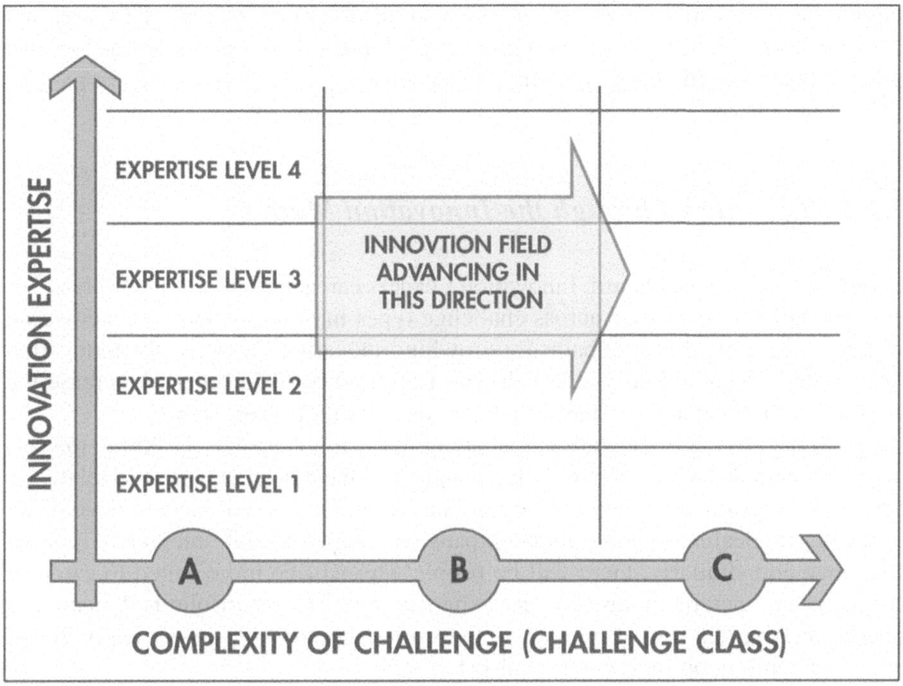

Fig. 3.4 The Innovation Matrix depicting Innovation Challenge Types versus levels of Innovation Expertise

Challenges". As one moves from left to right in the matrix, the nature of innovation shifts from more domain-specific skills-based work to a more strategic and integrative type of innovation.

The Y-Axis represents the expertise level through four levels, with Level 4 being the highest level of expertise. Each class of challenge involves its own ladder of expertise that practitioners climb as they gain more skill and expertise. Expertise in one class of challenge does not equate to expertise in another class of challenge. An innovator who is an expert at A could be at a novice level in a class B challenge, and vice versa.

The innovation matrix is one of the key instruments of the Innovation Leader. As stated earlier, the two main dimensions of influence that the Innovation Leader exercises are in (a) Amplifying the Impact and (b) Amplifying the Innovation Capacity of the System. The Innovation Matrix is a framework enables the Innovation Leader to track the organization's innovation as well as impact amplification capacities. The tool can also be used to assess and evaluate the organization's strategic goals against its innovation capacity.

Let us say that the Innovation Leaders are looking at their organization in terms of its Innovation Capacity. There are many factors other than just expertise and challenge type that make up the Innovation Capacity, such as organizational support for innovation, tolerance to exploration, tolerance to risk or failure, and tolerance to ambiguity (Surie and Hazy 2006); but for the purposes of this tool, we can roll them into the term "Innovation Capacity".

3.3.2 Pathways Through the Innovation Matrix

Using the tool as a dashboard, Innovation Leaders can get a sense of the "spread" or the "portfolio" of expertise across challenge types in order to shape an innovation strategy. There are different paths for growing Innovation Capacity; the Innovation Leaders have to judiciously select the most optimal paths in order to increase the total innovation capacity in their organizations and their ecosystem.

In a sense, for an organization, managing innovation capacity is akin to managing a diversified stock portfolio. Some people will have deep expertise in relatively narrow areas that are strategic for the organization. Others will have to develop an expertise in dealing with a more expansive, complex and integrated type of challenge and similarly, there will be people who will be more suited to carrying out the more operational tasks at any expertise level. This portfolio is dynamic, as people entering the lowest rung are able to move horizontally or vertically in this matrix depending on their career paths (Fig. 3.5).

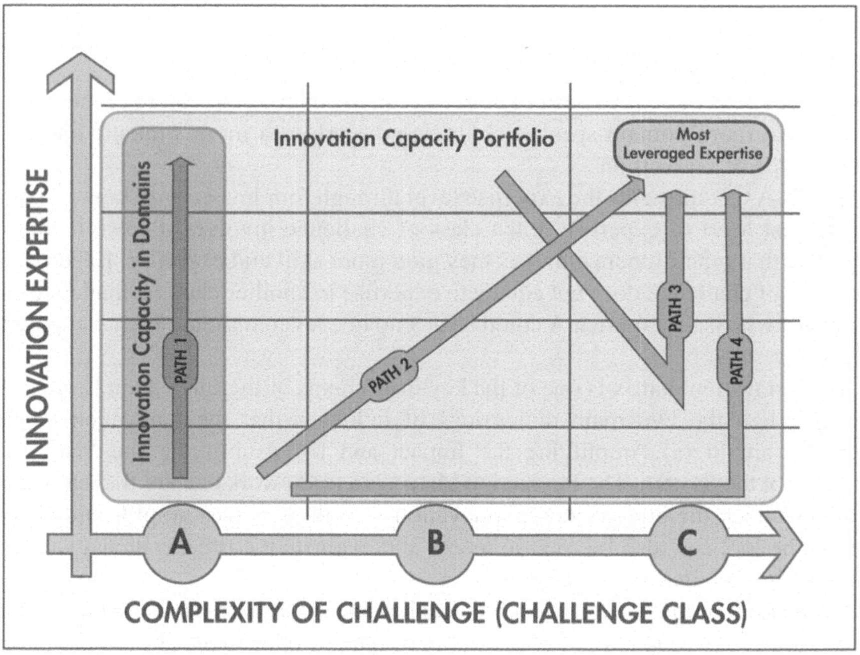

Fig. 3.5 Pathways in the innovation matrix

The figure above depicts four different paths:

- **PATH 1**
  This is a more traditional vertical growth in expertise in a given class of challenge. In this path, the individual's growth is achieved through the ability to carry out innovation with higher and higher levels of expertise, while the ability to handle challenges of higher complexity is not increased. For example, an organization with a product development division would naturally have an expertise ladder for someone entering as a fresh graduate out of an engineering or product design program and rising up the ranks to becoming an expert product designer.
- **PATH 2**
  This path increases the ability for an individual or organization to tackle more and more unbounded, complex and scaled challenges simultaneously with an increase in expertise levels. An example would be a curricular program whose students design gadgets and interfaces in the first year, and by the fourth year they are trained to innovate at the policy, organizational transformation, and platform architecture level.
- **PATH 3**
  This path illustrates someone who has already been performing at a high level of expertise in one class of challenge moving to a more complex regime. Expertise in one class does not translate to expertise in another class, so this path shows the person who initially operates at a diminished level of expertise as the challenge becomes wider, and then gradually regains a higher level of expertise. But an expert in any domain has the advantage of being intimate with the *notion of expertise*. She knows what constitutes expertise, knows what to do to get there, and knows how to translate heuristics from one domain into another. This journey tends to be much quicker if there is willingness and talent for the new class of challenge compared to growing expertise for the first time.
- **PATH 4**
  This path is undertaken by those who first go across the challenge types so that they understand the different regimes and then climb up the expertise ladder of a given problem class. This path implies a "rotation" across different domains without the expectation of high innovation expertise in any of them before choosing a specific domain, and then more innovation expertise within that domain. This is a good path for people who want to gain expertise in a specific area, but they want to have a career path that gives them flexibility and an ability to explore different domains as they grow.

3.4 Amplifying Innovation Capacity

In the previous section we have stated that the Innovation Leader is responsible for raising the innovation capacity of his organization or institution. But the question is: what constitutes an increase in innovation capacity? What are the skills, mindsets, capacities that constitute the various "Levels" and what attributes do we need to increase expertise as we move into handling more and more complex challenges?

This section illustrates the types of capabilities and skills that need to be developed or acquired in the expertise ladder. In the earlier section, we depicted multiple paths to enhanced Innovation Capacity. It is the responsibility of the Innovation Leader to increase Innovation Capacity through increasing the expertise level as well as the classes of challenges that their organization can address (A through C). As an illustration, we depict PATH 2 from the previous diagram – it is the path that moves diagonally across the innovation matrix, depicting a simultaneous increase in *both dimensions: innovation expertise and the challenge class.*

In Fig. 3.6, we identify four different levels of expertise, each associated with a particular set of skills, knowledge or attitudes. We considered and compared several alternative ways of characterization of innovation expertise for each level; eventually, we settled with the three dimensions along which the individuals will undergo transformation: (a) Skills, (b) Mindset, and (c) Impact frame.

An increase in expertise in innovation cannot circumnavigate an increased level of *Skills* in the various tasks involved in an innovation process. There are many skills to be mastered in innovation that require different cognitive modes (Christiaans and Dorst 1992) as the innovator becomes facile with the different stages of an innovation process, such as (a) primary and secondary research to understand the problem, (b) synthesis and framing to define the problem creatively and cast a visionary direction, (c) iteratively generating and testing concepts and prototypes with increasing levels of fidelity, (d) realizing and delivering tangible solutions that have the capability to make the desired impact, (e) amplifying the impact and building the capacity in the system to sustain the transformation.

In addition to a higher degree of facility, ease, and proficiency in the skills, it is also necessary for the individual to gain a *"meta-cognition"* of the skills and how

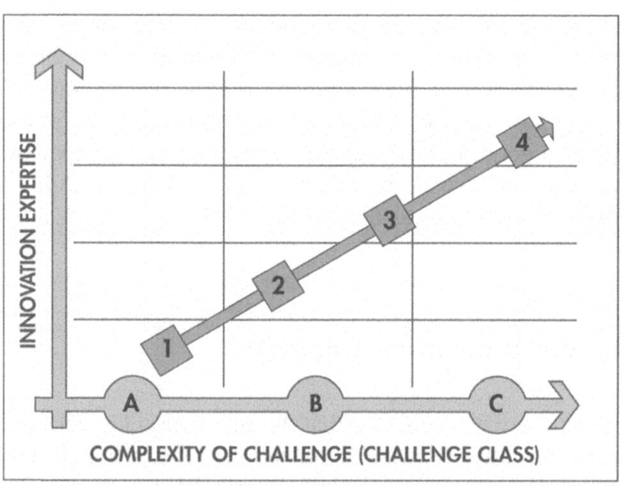

Fig. 3.6 For expertise levels along PATH 2 in the innovation matrix

they fit in the process, so that they can be leveraged effectively in diverse contexts. The question to pose is, "Are we seeing a change in levels in all the diverse skills that an innovator must master to become an expert?"

Alongside an increase in the skill level, an increase in the expertise and the ability to address challenges of greater complexity is not achieved without a change in *Mindset*. The innovator's mindset is comprised of dimensions such as his or her philosophical frame, worldview, value systems, ethical stance, creative confidence, agency, and overall attitude. At each level, the question to ask is, "Are we seeing the right shift in the mindset of the innovator?"

Finally, we consider the *Impact Frame*, by paying attention to what is the nature and scale of the ultimate impact that the innovator makes, and what is transformed as a result of activity. What is his sphere of influence - does he influence the outcome of a project or does he influence the behavior of an entire ecosystem? The question we are posing here is, "What is the nature and scale of the impact that that level of expertise is structured for?" (Fig. 3.7)

As individuals enter the innovation arena, their skills might be rudimentary in scope and sophistication, their mindset might be applicable to simple innovation tasks, they might balk at more complex challenges, and their potential for impact might be limited in scope. As they gain training, experience, and confidence, their abilities on each of these dimensions feed on each other and they are able to take on more and more challenging innovation tasks with greater and greater ease.

The following section will list the necessary attributes along the three dimensions for the various expertise levels along PATH 2 in the innovation matrix; in this path, the three dimensions of expertise grow simultaneously, as the innovator is simultaneously being exposed to greater and greater complexity of challenges as he gains expertise; skills, mindset and impact dimensions will not only have to increase in degree but also in scope.

3.4.1 Level 1 Expertise + Class A Challenges [Challenge Class A is the "Implementation or Adaptation Type Problems"] (Fig. 3.8)

Level 1 attributes:

(a) **Innovation skills**

i. Ability to carry out secondary research using multiple techniques.
ii. Analytical skills including numerical analysis and basic modeling of the broad scope of the challenge.
iii. Ability to carry out primary research and primary human factors' research using tools such as ethnography.
iv. Ability to ask the right questions.

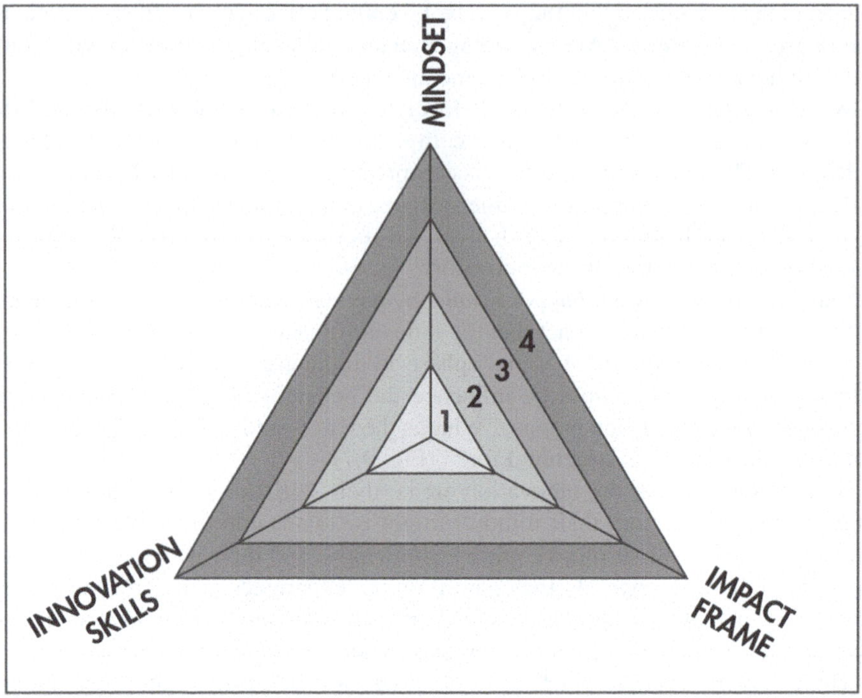

Fig. 3.7 A framework depicting the three dimensions that constitute an increase in Innovation Capacity

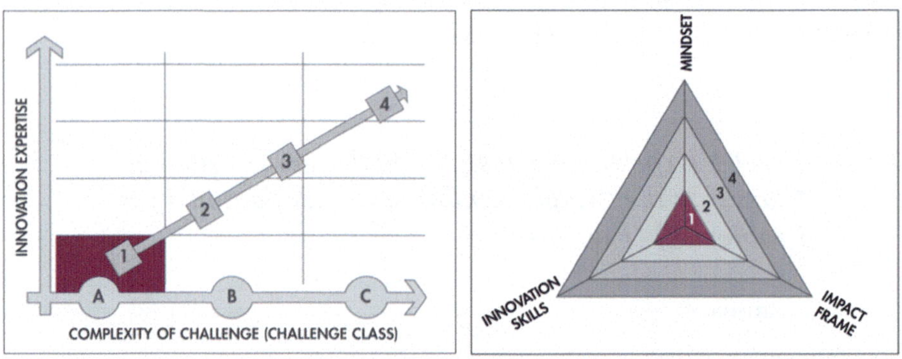

Fig. 3.8 Level 1 innovation expertise for class A challenges

v. Ability to synthesize qualitative and quantitative data and identify root causes.
vi. Ability to alternate between divergent and convergent modes of thinking.
vii. Given a frame and design principles, ability to generate highly creative concepts for product and service type challenges in the Class A category.

viii. Ability to communicate and develop concepts visually (Tversky 1969, 2005).
ix. Ability to fabricate and express concepts in a variety of media.
x. Ability to use prototyping as a way of thinking and designing.
xi. Ability to test prototypes with subjects while maintaining a keen eye for failure modes.
xii. Excellent teamworking skills and an ability to co-create with different types of experts.
xiii. Ability to communicate concepts visually, verbally, text media and video. Understanding of the power of storytelling and narrative structure.
xiv. Ability to take part in innovation discourse within innovation teams and with external partners.
xv. Ability to manage resources, relationships, and communication channels.

(b) **Mindset**

i. Quick to orient, curious, and a fast learner.
ii. Agile, flexible, and quick to pivot.
iii. Generative.
iv. Explorative.
v. Co-creative and collaborative.
vi. Confident but maintaining a "low ego".
vii. Sensitive to human needs.
viii. Hands-on.
ix. Cultured Naiveté – seeks to understand with a beginner's mind.
x. Bias to action.
xi. Self-driven.
xii. Optimistic.

(c) **Impact frame**

i. Influences project outcomes at each stage from Understanding to Delivery.
ii. Influences team spirit, energy, and dynamics.
iii. Influences speed of progress by moving rapidly.
iv. Influences the discourse by generating many questions, directions, and propositions.

3.4.2 Level 2 Expertise + Class B Challenges [Challenge Class B is "Open-Ended Medium-Scale Design Challenges"] (Fig. 3.9)

Level 2 attributes:

(a) **Innovation skills**

i. Ability to adhere to a high-level vision.
ii. Ability to shape research questions and direction of inquiry.

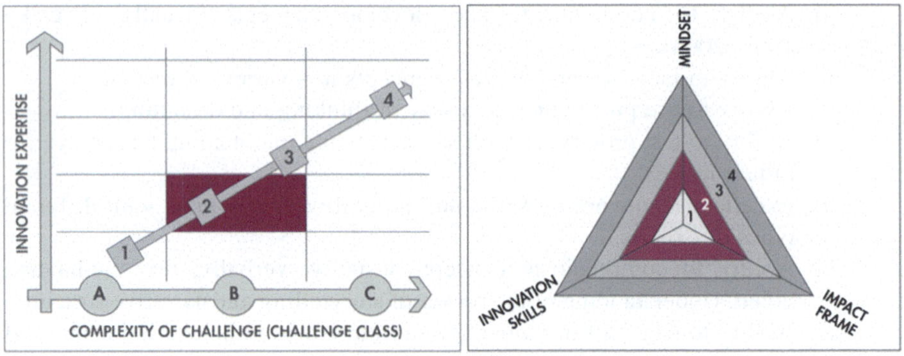

Fig. 3.9 Level 2 innovation expertise for class B challenges

iii. Ability to carry out critical thinking and exercise judgment in relatively ill-defined challenges.
iv. Ability to create incisive frameworks to depict the complexity of the problem.
v. Ability to carry out top-down, bottom-up, breadth-first and depth-first thinking.
vi. Ability to judge when to get the team to pivot from divergent to convergent and vice versa.
vii. Ability to think at an abstract level and then drill down to a detailed level.
viii. Ability to ensure, at each stage of the process, that many creative options are generated before down-selecting to the selected paths.
ix. Ability to keep many pathways open and avoid converging too early.
x. Ability to generate solution sets that meet multiple criteria.
xi. Ability to prototype more complex user interactions.
xii. Ability to develop a design to a high level of detail and completion.
xiii. Comfort with the innovation process and the ability to defend the process when it might be compromised.
xiv. Ability to mentor, direct and manage LEVEL 1 team members.
xv. Ability to manage projects with greater size, complexity, and number of moving parts.

(b) **Mindset**

i. Comfort with ambiguity and shifting boundary conditions of the problem.
ii. Comfort with reframing the brief.
iii. Comfort with revisiting assumptions if necessary.
iv. Comfort with facilitating co-creative activity with people more senior.
v. Comfort with being the advocate for the strategic direction.
vi. Comfort with leading a direction.
vii. Co-creative with a broader array of experts who might not be innovators.

viii. Managing relationships with peers, subordinates, leaders, partners, and external stakeholders.
ix. Process-minded and able to influence the process with shifting contexts and resources.
x. The ability to propose out-of-the box ideas and use the innovation process to mitigate the risks involved.

(c) **Impact frame**

i. Influences project framing and decision frame.
ii. Influences innovation direction and success criteria.
iii. Influences the discourse by engaging multiple stakeholders and facilitating a co-creative process.
iv. Influences leadership on direction, process, outcomes, and resource deployment.
v. Shapes and articulates the value proposition with nuanced level of sensitivity and critical thinking.
vi. Creates trusted relationships with internal and external stakeholders (Tidd and Bessant 2013).

3.4.3 Level 3 Expertise + Class B Challenges [Challenge Class B is "Open Ended Medium Scale Design Challenges"] (Fig. 3.10)

Level 3 attributes:

(a) **Innovation skills**

i. Ability to plan a research structure for an open-ended brief that can is efficient, and yet helps broad understanding.

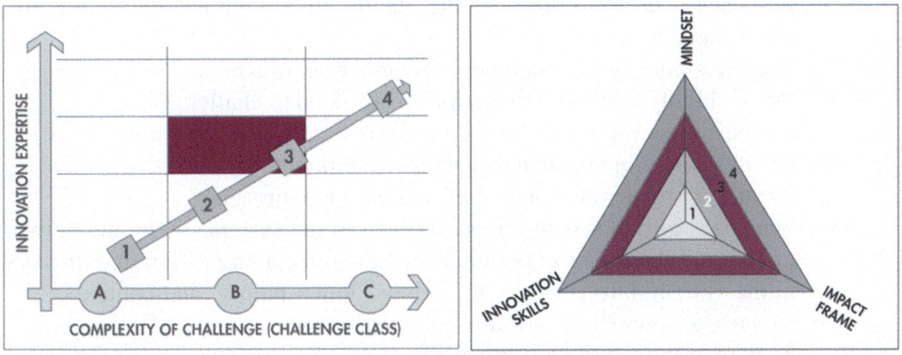

Fig. 3.10 Level 3 innovation expertise for more complex class B challenges

ii. Ability to take a holistic view of the challenge and delve into the hidden complexities of the system dynamics.
iii. Ability to craft and articulate a higher-level vision.
iv. Understanding the pathway of a user or stakeholder through the system.
v. Understanding organizational behavior.
vi. Understanding business opportunities and have the ability to generate innovative business models.
vii. Ability to identify causal pathways in more complex scenarios.
viii. Ability to join the dots and see patterns between seemingly disconnected parts of the system.
ix. Ability to communicate not just what the innovation is, but the underlying theory, the value proposition, the use cases, and why it is meaningful.
x. Ability to manage the innovation process all the way to deployment and beyond.
xi. Ability to manage an innovation team with diverse skills.
xii. Ability to build a measurement and evaluation protocol; ability to create dashboards for strategic decisions.

(b) **Mindset**

i. Thinks in terms of systems and sees things with a system designer's eye.
ii. Is perceptive of potential future failure modes in the system.
iii. Is sensitive to organizational culture and how to influence it.
iv. Is process-sensitive, and discerning of when the process is being compromised or subverted.
v. Values conceptual integrity and strength of platform architecture.
vi. Is improvisational and expands on thinking continually.
vii. Has the instinct to create the right enabling conditions for rich innovation.
viii. Values agility of mind – keeps him from falling into familiar ruts.
ix. Continually combines short-term value creation with long-term vision.

(c) **Impact frame**

i. Influences the entire journey of a user through a system.
ii. Influences the culture of the team, the organization, and the other stakeholders.
iii. Transfers knowledge, methodology and tools to others.
iv. Drives highly innovative solutions to ill-defined challenges.
v. Creates impact-mindedness in the entire organization or value chain.
vi. Creates a disproportionate return on investment.
vii. Creates a broader acceptance of innovation approaches.
viii. Sensitizes all the stakeholders involved to the systems view, the human factor, and the user's experience as they move along different pathways in the system, interacting with various "touch-points" and contexts.
ix. Creates a co-creative culture and engages the users and other stakeholders in the innovation process.

x. Builds in innovation capacity to manage the innovation journey beyond the project scope.

3.4.4 Level 4 Expertise + Class C Challenges [Challenge Class C is "Scaled Transformation Challenges"] (Fig. 3.11)

Level 4 attributes:

(a) **Innovation skills**

i. Ability to envision scaled transformations in the future and alternate system behavior.
ii. Ability to disrupt the prevalent norms and formulate alternate paradigms.
iii. Ability to trace causal chains in a complex "system of systems" type of contexts.
iv. Ability to analyze the challenge from a systems perspective with an understanding of feedback loops and relationships between the system "actors".
v. Ability to synthesize the findings across multiple layers of a system (social, economic, infrastructural, etc.).
vi. Ability to understand the human behavioral patterns not only from a needs perspective but from larger patterns of mindsets, culture, and motivational frames.
vii. Ability to understand multiple motivations of different system actors.
viii. Ability to facilitate innovation processes involving a genuinely trans-disciplinary engagement of members from different fields.
ix. Ability to design win-win propositions for stakeholders with different motivations.
x. Considers both the short-term and long-term ramifications.

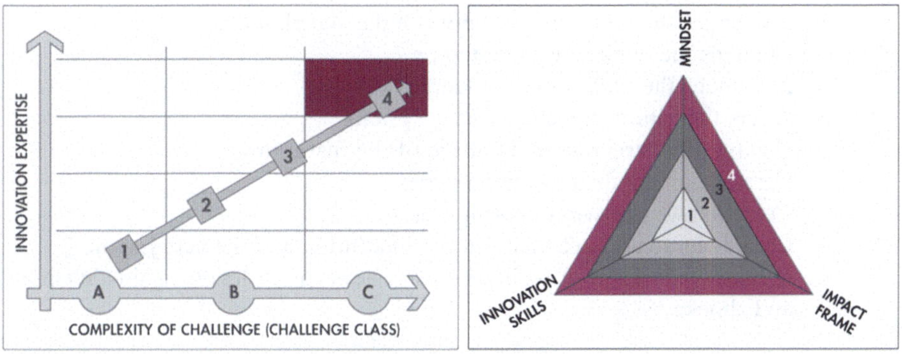

Fig. 3.11 Level 4 innovation expertise for class C challenges

xi. Innovates at the level of high-level strategies, institutional structures, governance structures, and policy.
xii. Builds coalitions among stakeholders and creates a co-creative environment.
xiii. Drives impact towards social and environmental needs.
xiv. Ability to generate intervention pathways that would transform system behavior, norms, and behavior.
xv. Garners political support, generates resources, and drives commitments for support, particularly for the scaling stages.

(b) **Mindset**

i. A continual quest for scaled impact
ii. Very aggressive drive for change in system behavior
iii. A mindset that weaves theoretical frames and practical processes.
iv. Belief in holistic, integrative, and interpretive thinking.
v. Transformation-minded.
vi. Continually seeking opportunities for win-win propositions in seemingly impossible motivational gridlocks.
vii. Confidence in challenging highly established systems and institutions.
viii. Thinks in terms of platforms and paradigms rather than products and services.
ix. Willingness to "en-rupt" systems and replace them with creative alternatives.
x. Deeply concerned about values, ethical nuances, and fair play.
xi. Mental clarity and conviction about intentionality and the philosophical stance.
xii. "Scale-mindedness" and a continual quest to increase the sphere of influence.

(c) **Impact frame**

i. Causes impact to the entire ecosystem and its behavior.
ii. Changes the paradigm within which the challenge is considered and the solutions are framed.
iii. Influences the relationships between the stakeholders.
iv. Changes the nature of the outcome.
v. Influences the scale of the outcome.
vi. Alters the future trajectory of the system.
vii. Changes the structure and culture of the institution.
viii. Creates new models and approaches.
ix. Creates new win-win opportunities.
x. Creates innovation capacity in the institution and the ecosystem.
xi. Increases the level of resilience of systems at all levels against changes and shocks.

3.5 Innovation Capacity as a Key Organizational Attribute

In the last century, the competency of an organization was measured in terms of its operational excellence and its ability to create a big market presence; in the twenty-first century, it is going to be measured in terms of its innovation capacity. The goal of most organizations is not just to deliver services, but to foster, change and improve lives (Drucker 2002). An organization that outperforms others, creates genuine impact, creates new markets, shifts paradigms, forges new directions, and solves the problems of the day is going to attract the best minds, will be the one that builds the strongest brand, and will be the one with the ability to withstand the turbulent winds of time. The ground is shifting from under the feet of organizations built around traditional models. Even in standard manufacturing and production settings, the emergence of complex production and systems (CoPS) is creating a need for new models and concepts for innovation since the current ones are based on high-volume consumer production (Hobday and Rush 1999).

An organization also has the choice of being a platform leader, one that drives sector-wide change, in which case there is a greater need for innovation (Cusumano 2002). An innovative organization is going give people the agency to innovate, provide value that is distinctive, give meaning to its own role in the marketplace, and set the tone for others to emulate. An organization with innovation capacity will celebrate the unknown, leverage failure, and understand the value of experimentation. It will find new ways of striking powerful partnerships with other organizations and with citizens. And the innovative organization will find ways of engaging with important challenges of the time.

Shifting from "Business As Usual" to having deeply held values around the power of innovation is not a simple shift. Changing organizational cultures will be one of the most critical and yet the most difficult tasks for an Innovation Leader. Some of the difficulty lies in the entrenched beliefs of the "Business As Usual". It is difficult for the existing leadership of any organization to acknowledge that the rules of the game are shifting rapidly or that they are not being innovative enough. The Innovation Leader's role therefore has an external function as well as an internal function. They must shape the nature of the external impact that the organization will have, and they must also shape the innovation culture of the organization, and in doing so they transform not only the DNA of the organization in which they belong, but also the larger ecosystem in which they operate.

If innovation capacity is already the marker of the avant-garde organization, and is going to be increasingly seen as the "table stakes" to compete, what are the implications for organizations that have not fully engaged in this critical dimension? The notion of innovation as a modality not only applies to an individual but also applies equally to a team, a project structure, a discipline, an organization, the mode of research, the decision structures, the value systems, and the modes of action. In short, it changes everything!

The implications are deep. It means that even before an existing organization figures out a new set of roles, organizational structures, Key Performance Indices, value systems, and altered bottom lines, it will need to find ways of increasing its innovation capacity. The path to this lies in *existing* members, at various levels, but most importantly at the top level, themselves adopting the role of Innovation Leaders. In a lot of cases, they will excel at operational excellence, business acumen, scholastic excellence, and general leadership, but might lack expertise in innovation expertise. It will be upon them to consider the innovation capacity portfolio of their organizations and their initiatives, and make the moves that would result in increased innovation capacity.

The first step for an existing leader is to endorse the value of innovation. It is important to initiate strategic innovation initiatives, to bring in innovative people, and frame innovation as a new dimension along which success is going to be measured. Existing members among an organization's leadership who adopt the role of Innovation Leaders will have to provide support and "executive air cover" for members who have displayed the talent for Innovation Leadership. They need to be empowered to explore, and be allowed to fail and propose directions that might be counter to conventional wisdom.

They will have to create "cultural enclaves" where people might have to go against the more prevalent company culture in order for them to feel safe in adopting innovation methodologies. The existing leadership will need to understand the need for innovation and will have to endorse new practices. The offices that handle the organization's strategic decisions will have to be most deeply involved in this transformation. Human Resources will have to use an entirely new lens to attract innovative people, incentivize innovation, reward innovative behavior, and train the entire workforce towards increased innovation capacity.

3.6 Conclusions: Implications for the Education Sector

For the education sector, there are even bigger implications. Education ought to be concerned with the shaping and preparing young minds for Innovation Leadership. Innovation and Innovation Leadership, instead of being another silo in the pedagogical system, need to be structured as a capacity across all disciplines.

But the educational institution as an organization, compared across other sectors such as industry and philanthropy, is not known for its agility and innovation. For the educational institution, in most cases, genuine innovation is not only an unfamiliar territory; it is often at odds with deeply entrenched institutional values. The University system celebrates scientific research in deep silos, publishing, and excellence in teaching, but does not tend to endorse creating innovative change agents who have a healthy disrespect for disciplinary silos. It claims to value multidisciplinary even when the reality is a very meager step towards genuine trans-disciplinary co-creation. There exists a lack of understanding of innovation methodology that giving institutions a false confidence in their level of innovation. The typical university not only fails to understand the value that innovation is going to have in the future, but many of the dominant value systems directly impede a culture of innovation.

Often, academic institutions, especially those in the lower echelon, tend to adopt a "trade school" approach, preparing students for the current market, placing disproportionate value on what has been already tried and tested. Academics all over also tend to clone themselves, and define success in terms of what they themselves have done, even though the global context of what academia ought to be has changed radically. The university has been slow to create a cadre of leaders who would meet the complex challenges with a more integrated and radically innovative set of approaches.

It is incumbent on the Education sector to shape young minds in preparation for careers that may not even exist today, and prepare then with the right skills, mindsets, and agency so that they can be active in shaping their own paths to making impact. The implications for educational institutions are twofold: they need to create an entire pedagogy around instilling innovation capacity in their students (so that they can go forth and exercise leadership in that dimension); but, more importantly, they need to build innovation capacity in their own organizations! The culture of the university is nothing but the summation of the value systems of the people running the organization. If they undervalue innovation, or, worse, still mistake what is being done as innovation, then it would be unrealistic for that university to develop that capacity in its students - and inevitably some other university with a better understanding would start attracting the best students.

References

Banerjee Banny (2009) Design: A field in Transition, Innovation, Journal of the Industrial Design Society of America, Winter 2008, Vol. 27 No. 4, 18

Banerjee Banny (2014) Innovating Large Scale Transformations, book chapter, "Design for Policy" edited by Christian Bason, Ashgate Publishing Company, Burlington, VT, USA, 71–86

Banny Banerjee, Designer as Agent of Change: A Vision for Catalyzing Rapid Change, Changing the Change Conference Proceedings, January 2008.

Brunswik E (1952) The Conceptual Framework of Psychology, International Encyclopedia of Unified Science (Vol. 1, No. 10, pp. IV + 102), Chicago: University of Chicago Press.

Bass BM (1990) From transactional to transformational leadership: Learning to share the vision. Organizational Dynamics 18(3):19–31

Christiaans HHCM and Dorst CH (1992) Cognitive Models in Industrial Design Engineering: A protocol study. Proceedings of the Fourth International Conference on Design Theory and Methodology. D.L. Taylot and L.A. Stauffer, New York, ASME Press: 131–137.

Cross N (2006) Designerly Ways of Knowing, Springer-Verlag, London: 77–89

Cross N, Christiaans HHCM, Dorst K (1994) Design expertise among student designers. Journal of Art and Design Education 13:39–56

Drucker PF (2002) The Discipline of Innovation. Harvard Business Review 63(3):67–72

Dweck CS (2008) Mindset: The New Psychology of Success. Random House, New York

Findeli A (2001) Rethinking Design Education for the 21st Century: Theoretical, Methodological, and Ethical Discussion, Design Issues: Volume 17, Number 1 Winter 2001: 5–17

Cusumano M, Gawer A (2002) The Elements of Platform Leadership, MIT Sloan management review, ISSN 1532–9194, Vol. 43, N° 3 , págs. 51–58.
Goldstein J, Hazy JK, Lichtenstein B (2010) Complexity and the Nexus of Leadership: leveraging nonlinear science to create ecologies of innovation. Palgrave Macmillan, Englewood Cliffs
Hazy JK (2007) Computer models of leadership: Foundation for a new discipline or meaningless diversion? The Leadership Quarterly 18(4):391–410
Hammond KR (1980) Introduction to Brunswikian theory and methods. New Directions for Methodology of Social and Behavioral Science, No. 3, 1–11.
Hobday M, Rush H (1999) Technology management in complex product systems (CoPS) – ten questions answered, International Journal of Technology Management, Volume 17, No. 6, 1741–5276
King N, Anderson N (1990) Innovation in working groups. In M. A. West, & J. L. Farr (Eds.), Innovation and creativity at work, 81–100: Psychological and organizational strategies. Oxford, England: John Wiley & Sons, xiii, 349.
Mumford MD, Scott GM, Gaddis B, Strange JM (2002) Leading creative people: Orchestrating expertise and relationships. The Leadership Quarterly 13:705–750
Nelson RR, Winter SG (1982) An Evolutionary Theory of the Economic Change. Belknap Press, Cambridge MA
Rosing K, Frese M, Bausch A (2011) Explaining the heterogeneity of the leadership-innovation relationship: Ambidextrous leadership, The Leadership Quarterly, Volume 22, Issue 5: 956–974.
Sarros JC, Cooper BK, Santora JC (2008) Building a climate for innovation through transformational leadership and organizational culture. Journal of Leadership & Organizational Studies 15:145–158
Shipton H, Fay D, West MA, Patterson M, Bird K (2005) Managing people to promote innovation. Creativity and Innovation Management 14(118–128):2005
Somech A (2006) The Effects of Leadership Style and Team Process on Performance and Innovation in Functionally Heterogeneous Teams, *Journal of Management, February 2006, 32*: 1, *132–157.*
Stogdill RM (1974) Handbook of leadership: A survey of theory and research., New York, NY, US: Free Press, 613.
Stogdill RM, Coons AE (1957) Oxford, England: Ohio State Univer., Bureau of Busin Leader behavior: Its description and measurement. xv 168
Surie G, Hazy JK (2006) Generative Leadership: Nurturing innovation in complex environments, Emergence: Complexity and Organization, Issue Vol. 8 No. 4 2006: 13–26
Tversky B (2005) Image and Reasoning, Keio University Press, Tokyo, 15–29
Tversky B (1969) Pictorial and verbal encoding in a short-term memory task. Perception & Psychophysics 5:225–233
Tidd, J, Bessant J (2013) Managing Innovation, Wiley; 5 edition, New York, 299–319
Uhl-Bien M, Marion R, McKelvey B (August 2007) Complexity Leadership Theory: Shifting leadership from the industrial age to the knowledge era, The Leadership Quarterly, Volume 18, Issue 4, 298–318
Valkenburg AC, Dorst K (1998) The reflective practice of design teams'. Design Studies 19(3): 249–271

Chapter 4
A New Kind of University

Randy Swearer, Véronique Hillen, and Paola Bertola

4.1 Introduction

In the previous chapters, we have critically examined the reasons behind the current cross-sector demand for innovation. We have also outlined the nature and scales of the challenges innovation leaders confront as they navigate morphing economic and social conditions characterized by complexity, ambiguity and disequilibrium. In order to succeed in these conditions, the innovation leader uses strategies focusing on framing problems in complex systems, co-creating in transdisciplinary contexts, and scaling solutions.

This chapter explores the mismatch between the education universities are designed to provide and the education innovation leaders need to succeed. The concepts developed in this section are built on in Chap. 5 and chapters in Part II, which concretely explores how individual university innovation programs throughout the world are experimenting with new modes of learning, and in the process becoming laboratories for rethinking the meaning, purpose and organization of universities.

Universities have enormous roles to play as societies around the globe face social, economic, and environmental challenges of enormous complexity. In order to create a new generation of innovation leaders to confront these challenges universities will have to undergo a profound structural and cultural change affecting their missions, values, literacies, and broader ecosystems. We have no illusions

R. Swearer (✉)
Design Dept. Politecnico di Milano, via Durando 38/a, 20158 Milano, Italy
e-mail: swearerr@philau.edu

V. Hillen
Paris Est d.school, 6 avenue Blaise Pascale, 77455 Champs sur Marnes cedex, France

P. Bertola
Philadelphia University, 4201 Henry Avenue, Philadelphia, PA 19144, USA

B. Banerjee, S. Ceri (eds.), *Creating Innovation Leaders*,
Understanding Innovation, DOI 10.1007/978-3-319-20520-5_4

about how challenging this transformation will be for our systems of higher education. Yet, contrary to the popular stereotypes about the inability of universities to change, we see many encouraging academic innovation initiatives around the globe. Our broader project beyond this chapter is to disprove the caricatures of the change-resistant university by concretely applying at larger scales what we have learned from these projects. An analysis of these emerging programs will be carried out in the next chapter.

4.2 A New Paradigm for Creating Innovation Leaders

Forces Driving Systemic Change in Education

The world our graduates are entering has become progressively more dynamic and transient due to macroeconomic trends resulting in interdependent global economies that are continually restructuring. This condition is driven by a variety of factors, including increasingly fluid capital markets, new digital platforms for collaborating and communicating that virtually distribute human presence, rapid cycles of technological disruption, and general globalization trends. Those organizations succeeding by quickly pivoting toward new opportunities have replaced the path dependencies of industrial era institutions and businesses. These conditions have accelerated the rate at which professions are born, die, and evolve. Our educational models must deeply respond to these trends.

Twentieth-century organizational models persist, but they were not designed to address the challenges presented by these conditions. Their hierarchical and siloed structures were organized to segment problems into more manageable fragments, but they were much less effective at dealing with the kinds of ill-defined, systemic problems we face today. As discussed in previous chapters, *wicked problems* are messier and more ambiguous in nature, more connected to other problems, more likely to react to solutions in unpredictable nonlinear ways, and tend to produce a high frequency of unintended externalities. For these reasons, many of our long-standing institutions, including universities, are now struggling to adapt.

Innovation-driven organizations operating in this new environment expect university graduates to conceive of, develop and implement real solutions to ill-defined problems in order to facilitate positive change. In order to meet these expectations students must be more entrepreneurial, take more active roles in their work, be agile thinkers, integrate disciplinary frameworks, collaborate across functions, manage creative teams, develop context-sensitive solutions, and have an expanded critical awareness of the world in which they live and work. In other words, they must expect the kinds of agency and awareness typically associated with innovation leadership.

4.3 The Transformation Imperative: How Higher Education Must Change

Becoming Critically Aware of the Historical Model We Have Inherited

The project of reimagining our universities in ways that foster innovation leadership will be one of the major challenges our societies face in the coming decades. A helpful first step toward confronting this challenge is to become critically aware of the historical model we have inherited.

Universities are among our oldest and most enduring institutions. They are medieval in origin, but were reimagined during the Enlightenment to serve the industrial era. They emphasize machine age ideals: efficiency, hierarchy, standardization, punctuality, quantitatively measurable productivity, scientific management, and the compartmentalization of knowledge. Between the late nineteenth century and the first quarter of the twentieth century dozens of functions specifically created to serve the industrial era were invented and institutionalized in higher education; for example, the template for research universities, standardized guidelines for accreditation, graduate schools serving clearly delineated professions (business, law, nursing, etc.), grading and multiple-choice tests, the Carnegie unit of credit, and highly structured degree requirements (see Davidson, 2011).

We need to become critically aware of this inheritance in order to design new *post-industrial* university models that serve what Robert Darnton (2008) has termed the *fourth age of information*. This era, which began with the spread of the Internet in the early 1990s, has fundamentally changed our politics, social arrangements and economic conditions. Its effects on us have been pervasive, yet our systems of higher education remain organized by legacies from the industrial era. Consider the purpose of professional knowledge in the industrial era and contrast it with economic conditions today. Professional knowledge in the industrial age focused primarily on the process of converting raw commodities into the production of standardized tangible goods with specific feature sets for users. This ultimate goal conditioned the way universities organized themselves into delivering professional education.

Yet today substantial sectors of our economies produce value in radically new ways. For example, in advanced economies value is increasingly generated by using goods (mostly produced elsewhere) as props, and services as a stage, to create memorable *experiences* for customers (Pine & Gilmore, 1998). This new paradigm inverts many of the central characteristics of the industrial economy. Goods become experiences, manufacturing becomes staging, standardization becomes personalization, users become guests, features become sensations (Pine & Gilmore, 1998). Universities have, however, been slow to consider the implications of changes in value creation such as these both in the area of professional education and their own organizational structures. What would an *experience university* look like, and how would it create integrative, memorable, personalized learning experiences for its students?

4.3.1 Legacy University Models Cannot Provide What Innovation Leaders Need

Consider the machine-age inheritance of universities and the contradictions it creates for us today. Universities have organized themselves into increasingly specialized silos of knowledge, guarded and assessed by professional organizations and accreditors, but innovation leaders must be able to think in ways that are integrative and transdisciplinary, and act in ways that are appropriate and make sense for a given context and users. Knowledge has been taught with standardized hierarchical pedagogies reflecting labor arrangements of the industrial era, while innovation leaders need to learn in more egalitarian and collaborative environments. The epistemologies of the industrial era university have emphasized inductive and deductive reasoning, while the innovation leader must reason by fluidly shifting among induction, deduction, and abduction. Across many dimensions such as these our current universities have organizational models and epistemological orientations that simply do not, and cannot, nurture the innovation leadership that is crucial to solving the wicked problems of our day.

Perhaps one of the most striking shifts is the change in the role of the professor from knowledge disseminator in one discipline to multidisciplinary stage director facilitating intentional action situated in specific contexts (Hillen & Levy, 2013). This new role positions teaching as a transformational experience for both faculty and students. Defined and controlled classroom topics are replaced by real-world problems characterized by messiness and unpredictability. Knowledge continues to be essential but its translation into know-how (Dewey, 1925) drives learning. The boundaries around physical learning spaces become porous membranes through which students and faculty pass as they move to and from experiences in the real world of people and practices.

In fact, learning experiences that are crucial for creating innovation leaders involve identifying the subject of study from the plethora of variables students and faculty confront in the field. These variables are entangled in the specific, messy, and complex circumstances that define the countless facets of human experience.

Understood in terms of courses, these learning experiences normally begin with a thematic framework, but students and faculty formulate the actual subject by determining the primary problems faced by individuals and groups in specific contexts. For example, an entrepreneurship course might have the general theme of aging at home, but during the course of research, students might discover that a more salient problem for their course concerns how older people transition to navigating the aging process in senior communities with a range of assistive services and social supports. The subject of student work might become looking for value creation opportunities that support the phased transition between the social, cultural, economic, and health-support frameworks of the private home and those associated with life in a more communal setting. Subjects such as this are truly wicked: indeterminate, without clear boundaries, and connected in systemic ways to other complex problems.

From Flipped Classrooms to Flipped Universities

The new learning-centered university environment consists of easily accessible, pervasive digital content. Much has been written about the fact that this condition fundamentally changes the role of instructor from primary content provider to co-navigator of the learning process in the additional roles of mentor, coach, and knowledge curator. In this new world the instructor adds value not by providing commoditized content to students, but by adding unique integrative value to the learning and knowledge-building process.

The well-known *flipped classroom* teaching trend responds to this new world of pervasively available content by asking students to learn content outside of class through multiple delivery channels, including books, digital databases, online video tutorials, Massive Open Online Courses (MOOCs). Time with the instructor is used for active forms of learning: synthesizing information into knowledge and applying it in various contexts. The flipped classroom is an important first step in developing (and popularizing) classroom-level pedagogical responses to our new information-rich learning environments, but we would like to see the same concept scaled to the university level. How could a *flipped university* create new kinds of learning ecosystems that advanced the necessary values and literacies for creating innovation leaders?

4.3.2 Enlarging the Definition of Building Knowledge

Orienting teaching toward complex problems, know-how, action, and co-creation is in many ways the opposite of how most academics have been taught to think, teach, and find opportunities for recognition and advancement in their fields. The academy has typically approached complexity with reductionist methods that define and tackle problems by constraining their scope.

This process of constraining typically involves narrowing areas of inquiry to specific subjects that academics study in depth. The strategy is to study the parts of a system to build a general body of principles about its universal properties. A literature scholar might specialize in Jane Austen, placing her work in a broader system of scholarship about gender roles in works by female writers. A molecular biologist might specialize in protein biosynthesis of frog embryos to identify general principles regarding the more general properties of cellular systems. These principles would then be in turn deductively applied to understanding the nature of the system components. Of course, we do not dispute the value and profound contributions to knowledge made by this kind of work. However, we argue that other approaches to building knowledge that fully account for the complex lived systems we inhabit have been marginalized, as has the invaluable knowledge derived from reflective practice (Schön, 1983).

4.3.3 Becoming Critically Aware of Major Differences in Education Models

We have summarized several differences between our industrial-age universities and what is needed to educate innovation leaders. The following table contrasts the differences between the traditional university model and a new model capable of deeply integrating the literacies of innovation (Table 4.1).

As this table demonstrates, the chasm is wide between our industrial-era universities and a new model for higher education that would broadly support innovation. Yet there is reason for hope. The next chapter will critically review academic experiments on many continents that are changing the ways we think about what an innovation-driven university education means and how it could be effectively delivered. In order to create a wider lens for viewing these programs, we imagine a new kind of university based on fresh assumptions. We frame this new kind of university as a *platform* built to deliver particular values, missions, goals, and organizational models linked to innovation leadership (see Sect. 4.2).

4.4 Imagining a New Kind of University

Universities seeking to educate innovation leaders face two levels of challenges. The first level concerns the learning environment for students: its structure, orientation, pedagogies, course structures, and academic programs. These issues are concretely addressed in Chap. 5 and Part II. Yet in order for these experiments to scale, the university itself must change. The second level concerns the university model: its mission, values, and literacies

Platform Learning

We use the word *platform* as a conceptual model for an evolving university model driven by a focused mission and new values that support innovation literacies and their supporting assumptions. A platform allows us to perform a set of actions that take place beyond it. It is "outward facing" and facilitates our agency in the world. The strength of a platform should be measured by the extent to which it enables us to do the things we want to accomplish with as little presence and resistance as possible. In fact, the ideal platform, if it existed, would be transparent. A well-designed smartphone is a concrete example of a platform. It brings together a software framework and hardware architecture to create a platform for doing things beyond it with a minimum of intrusiveness (Fig. 4.1).

Table 4.1 Traditional university model vs. a new university model incorporating innovation literacies

Assumptions of the traditional university model	Assumptions of a new university model incorporating innovation literacies
From knowledge-centered to problem-centered	
Knowledge should be generated by pursuing clearly delineated research questions in specialized knowledge domains.	Knowledge should be generated in an iterative process of framing ambiguous problems in transdisciplinary settings.
The world around us is stable, allowing us to reliably find and solve problems by using the past to predict the future.	The world around us is rapidly transforming, requiring us to find and solve problems by supplementing or replacing predictive modeling with building scenarios and tangible solutions that evolve in iterative cycles.
Systems thinking should be discipline- specific.	Given the complexity of many wicked problems and the imperative that students integrate disparate sources of knowledge to find and solve them, systems thinking should be a basic literacy requirement for all students.
From bounded to porous	
The university should be a self-contained institution. It provides within its boundaries access to most kinds of knowledge, experts and learning infrastructures needed by students.	The university should be porous and designed to create a dynamic balance with external agents, such as generalists, stakeholders, experts, institutions, governments, companies, etc. The focus is on accessing and integrating the best and most appropriate sources of knowledge to find and solve problems—inside or outside of the university.
From experts to knowledge integrators	
Disciplinary expertise should be the ultimate goal of learning	Disciplinary expertise should be valued, but it is equally important to integrate disparate kinds of knowledge to find and solve problems.
Disciplines should be clearly bounded and defended. Knowledge should be assumed to have a relatively long shelf life.	Students should be taught to understand disciplines as existing on a continuum with overlapping boundaries. This continuum is always in a state of flux.
From knowledge to learning	
Because knowledge should be assumed to be relatively stable, learning occurs primarily during a student's experience at the university.	Because knowledge should be assumed to be fluid and rapidly evolving students are taught to learn throughout their lives. The university becomes a platform supporting this process.
From passive to active learning	
The university should be teaching centric. Disciplinary expertise should be the ultimate goal of learning. The expert instructor delivers knowledge to the novice student.	The university should be learning centric. Disciplinary expertise should be valued, but equally important is the student's ability to find and solve problems by actively integrating many kinds of knowledge from disparate sources. The instructor guides the student during the knowledge integration process.

(continued)

Table 4.1 (continued)

Assumptions of the traditional university model	Assumptions of a new university model incorporating innovation literacies
Instrumentality should be discounted as vocational learning	Instrumentality should be embraced as a form agency and a means of creating knowledge.
From assessing domain expertise to evaluating how groups perform with real-world projects	.
Learning should be competitive	Learning should be cooperative
Assessment focuses exclusively on the student's mastery of formal, hard knowledge.	Assessment includes soft knowledge, such as collaboration and teamwork, decision-making, and leadership.
Soft knowledge is discounted, hard knowledge is highly valued	The entire continuum of hard and soft knowledge is embraced.
Assessment should focus on students' explicit disciplinary knowledge. There are right and wrong answers. Faculty member alone conducts assessment.	Assessment incorporates student's facility with creating, applying and improvising with multi-disciplinary knowledge in disparate contexts. There are appropriate solutions, which are discovered during the learning process. The impact of the project on the real world is the focus of assessment. Therefore, multiple stakeholders inform the assessment process.

Fig. 4.1 University as a platform

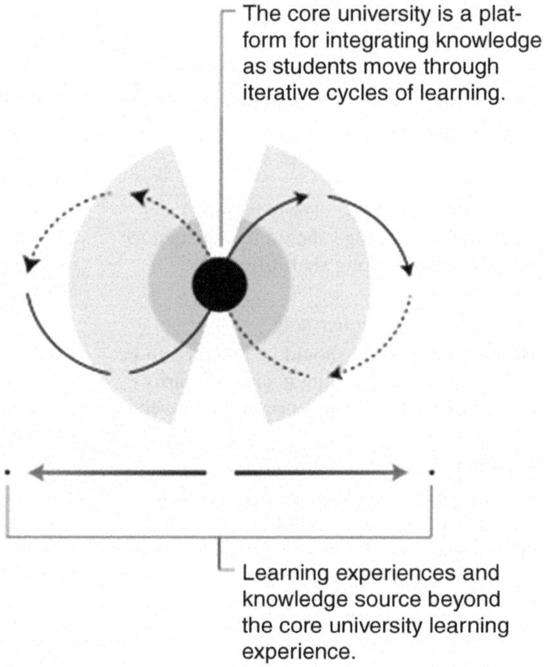

4.4.1 The Innovation-Driven University: New Mission and Values

A common new mission for the platform university is to create innovation leaders to focus on addressing the classes of problems identified in Chap. 1 (problem classes A-C and their subtypes).

The Value of Student-Centered Learning

In order to pursue this mission our platform university *would value a learning environment that is radically student-centered.* The university platform would allow students to access all of the learning and infrastructure they required in order *to focus on addressing the problems they pursued.* The roles of the faculty, staff and administration would be to facilitate the abilities of students to learn in ways that enhanced their agency—and then get out of the way. This facilitation would involve three pillars: teaching in new ways that emphasize staging the learning experience, mentoring and coaching, brokering knowledge, and advising; working with students in human-centered transdisciplines; creating an ecosystem of relevant actors (e.g. outside partners, content experts, faculty); and staging the learning process with fieldwork and studio (or studio-like) active learning experiences (see diagram above).

The Value of Personal Responsibility

What would learning look like if the course platform deeply embedded the value of personal responsibility for the state of the world? The implications would be vast. The content of the course would focus on the knowledge, skills, habits of mind, and tools students needed to imagine improved future states and to proactively design paths toward them. Instructors and students would be active agents for positive, transformative change.

The Value of Integrating Knowledge to Act in the World

The most fundamental locus of learning at universities is typically the course. It is a microcosm where the values of the broader university are experienced as learning. If we reimagine these values so that they support innovation, how would the course model change—and what would this transformation imply about the broader university?

If a course became a platform, it would be designed from the ground up to facilitate and leverage student efforts to access and integrate disparate sources of knowledge for the purpose of acting. Students would need to learn on demand and sort out what happens when general knowledge collides with situated practice. *It would embed the value of collaboratively integrating knowing and acting to improve the world.* The knowledge required to do so would often be multi- or trans-disciplinary and would be accessed by students interacting with a network of distributed resources: the communities that anchor their work, other universities, cultural and scientific institutions, media, outside content experts, databases, mentors, MOOCs, and so on. The course platform would at every level stress the importance of agency: student's abilities to effectively and thoughtfully act in world by exploring, accessing, assessing, analyzing and synthesizing knowledge.

The Value of Diversity for Collaboration and Co-creation
The knowledge an individual student brought with him or her into the world would often be insufficient to act in ways that enabled him or her to address wicked problems. Effective action directed toward solving complex problems often requires intellectual, political, and cultural diversity of a group rather than the capabilities and background of an individual. Therefore, platform learning would embed the value of diversity for collaborative and co-creative learning.

The values we have described that drive platform learning, integration for action, collaboration and co-creation, and student-centric learning are all oriented toward creating innovation leaders with a common set of agilities. Although we focus here primarily on students and faculty, we believe that in order deeply embed platform learning it is imperative that our universities incorporate the fundamental innovation literacies *at all levels of their internal organizations,* including administrative staff, academic support offices, academic program managers, executive leaders, and trustees.

Generating Value in the New Learning Ecosystem
Students live in rich, information-dense environments, which shifts the ways universities create value for their students. Instead of primarily providing widely available commodity content to students, universities must shift to providing fertile learning environments for integrating information to build knowledge. The bounded classroom conceptually transforms into a workshop or studio with porous boundaries, where students work with faculty to integrate learning from the field, online sources, books, co-curricular activities, etc. It is about progressively iterating ideas and assessing them with peer input, faculty guidance, and often experts from other disciplines or fields beyond the campus boundaries. Learning becomes a self-reflective creative process of building knowledge and applying it in iterative loops of learning.

There are over 9000 diverse universities in the world. With the burden of content delivery no longer a primary concern, universities have the unprecedented opportunity to differentiate themselves based on the unique integrative value they bring to the learning process beyond delivering commoditized content. The following equation describes this critical new mode of value creation:

Student university experience minus commodity content
= added value

4.5 Leadership and the Literacies of Innovation

Innovation leadership exists on a continuum organized in Chap. 1 with four points of reference (Levels 1–4). We want to be clear that in academia this nomenclature *does not necessarily imply a rigid hierarchy of functions and power relations.* The values we outline above anchor the learning process and center it on the ability of all innovation leaders on this continuum to develop *modes of agility:* working across sectors and disciplines, integrating disparate forms of knowledge and reasoning, co-creating in a wide range of collaborative contexts, and learning and acting by building knowledge with things. We understand these agilities as *innovation literacies,* in the sense that innovation leaders must be fluent with them in order to address the complex challenges they face. While it is true that innovation leaders at different organizational levels will use the literacies in distinct ways, all must understand the ways literacies interact to create a dynamic framework for innovation.

Innovation Literacy 1: Agility of Collaboration
All types of innovation leaders collaborate in diverse working environments. Collaboration is essential because the kind of value innovation leaders generate is typically co-created by cross-sector teams. These teams are structured with relatively flat hierarchies to reduce the chances that power relations among members will impede the exchange and integration of knowledge. Innovation leaders view the process of co-creating in teams as instrumental and developmental. We typically view co-creation in instrumental terms: a group of collaborators work together to further the goals of a project. Yet, innovation leaders also understand collaboration as a formative learning process. The project becomes a common social *learning medium* through which team members exchange and integrate forms of formal and tacit knowledge.

Two crucial forms of interaction for successful co-creating on teams are *bonding* and *bridging* with others (Arthur, Defillippi, & Lindsay, 2008). Bonding relationships among co-creators involve sharing related pools of knowledge within similar professional cultures. While innovation leaders must be adept at bonding, it is especially important that they are comfortable with bridging gaps in the co-creation process by forming strong relationships with individuals from dissimilar knowledge domains and professional cultures; for example, a software engineer might bridge a gap on a project team by forming a relationship with a user experience designer.

Studies of IDEO's designers (A. Hargadon & Sutton, 1997) and Thomas Edison's invention dynamics (A. B. Hargadon, 2002) have helped to demystify the cult of the inventor and demonstrate the importance of bridging and bonding as a foundation for innovating. Using the example of Thomas Edison, from 1876 to 1881 his laboratory in Menlo Park, New Jersey, created over 400 patents and was known as an invention factory. It produced innovations in high-speed telegraphs, telephones, phonographs, generators, mimeographs, light bulbs, and much more. Edison and his team were knowledge brokers: They learned by bridging and bonding not only within the team, but far beyond it with experts in knowledge domains much different than theirs. This constantly shifting network of new connections fueled one of the most innovative companies of the late nineteenth and early twentieth centuries.

Innovation Literacy 2: Agility of Function
Innovation leaders in organizations must have the mental agility to improvise with organizational functions by moving between them. In order to do so they need the ability to mentally frame the organization's internal dynamics as a whole, and understand how individual functions add value to one another. Using our typology of innovation leaders, this ability to broadly frame functional dynamics in organizations is especially important for *level 3*. They must broadly developmental models of an organization's functional units in order to envision creative ways of combining—or transforming—them. They see functions as overlapping on a continuum. In order to build these mental models and constantly update them, organizations emphasizing innovation tend to be functionally transparent.

Innovation Literacy 3: Agility of Thought
Innovation leaders understand a simple truth: integrating dissimilar kinds of knowledge is the engine of innovation breakthroughs (Berggren, 2011). This literacy is so crucial for innovation leadership that we discuss it more broadly here.

We can no longer sift knowledge into comfortable categories in an age where it is generated and evolves so rapidly and is applied so broadly. Theorist George Siemens (2006) captures this concept: "We must learn to dance (engage and interact) with knowledge in order to know what it is" (p. 20). The innovation leader is adept at integrating contrasting sources of knowledge and, depending on his or her position in the organization, creating the conditions that encourage knowledge integration. Universities must respond to the need for knowledge integration by developing layered academic programming that puts the content and inquiry methods of single disciplinary domains in dynamic relationships with each other, but also links them together with human-centered trans-disciplines.

Harvard's School of Engineering and Applied Sciences (SEAS)
Harvard's School of Engineering and Applied Sciences (SEAS) has integrated this concept into the core of its mission. Since the inception of Harvard's newest school, SEAS leadership has recognized both the opportunity and challenge of preparing its graduates with new methodologies and processes to solve the complex challenges of the twenty-first century. It has positioned the school to create engineers who are literate in liberal arts as well as to help non-engineers throughout the university increase their engineering literacy. This commitment to fostering dynamic opportunities to learn and engage across disciplines has led to a number of new programs and courses.

(continued)

Established in 2007, SEAS has no departments, mostly interdisciplinary research and a substantial portion of cross-disciplinary and system-level courses that are transforming undergraduate engineering education. The SEAS approach to multi-disciplinary design, engineering and entrepreneurship has created significant student demand and growth in the faculty, facilities, and staff.

Engineering embedded in a liberal arts framework provides unique opportunities, especially when innovation-focused education is integrated throughout the curriculum. Two of Harvard's most diverse team and project-based experiential learning courses grew out of SEAS, The Innovators' Practice and Design for Desirability. Both courses, developed by Dr. Beth Altringer, are now cross-listed in the Graduate School of Design (GSD). In four years, they have attracted students from nearly every discipline across the university. Building on initial successes, SEAS and the GSD are developing a masters-level interdisciplinary program in 2015.

The highly specialized training and acculturation of the professorate in single disciplines often creates the conditions for conflict during multi-disciplinary collaborations as each participant seeks to assert the framework of his or her discipline's inquiry methods. Here, agility of thought helps professors and students to recognize a broad range of inquiry methods that drive disciplinary thinking. They do not need to master these inquiry methods, but they should understand the kinds of questions pursued by each discipline or discipline cluster. This recognition creates the conditions for innovation by defining a conceptual space for integrating disparate forms of knowledge on transdisciplinary teams.

Alta Scuola Politecnica

The Alta Scuola Politecnica focuses specifically on the issue of multi- and trans-disciplinary inquiry methods. It is a joint program between Politecnico di Milano and Politecnico di Torino in Italy. Since its inception it was targeted at a diverse student community representing all branches of engineering, architecture, and design. The program offers gifted students the opportunity to have a transdisciplinary educational experience in parallel with their discipline-based Masters programs. It focuses on developing cognitive processes that are central to innovation.

The Alta Scuola Politecnica is based on field projects tutored by a multidisciplinary faculty team that focuses on real cases submitted by external partners, such as companies and private and public institutions. Commonly, projects are related to problem-setting within complex, system-level contexts where the innovation process could play a substantial role. Students are encouraged to find dynamic relationships between their own discipline-

(continued)

based methods for modeling problems and those of their teammates. This learning environment creates a space for reframing the original, assigned problem to discover more fundamental questions that could drive the innovation process. Problem setting, concept generation, and scenario design are the core of field project activities, focusing on two important goals. The first goal is to offer external partners a fundamentally unique perspective on innovation strategies and possible scenarios that would not typically emerge from a traditional R&D environment. A second goal is to expose students to different inquiry methods and problem-modeling systems to encourage lateral thinking and sense-making among very heterogeneous bodies of knowledge.

Framing Academic Inquiry

In his seminal book, Nigel Cross (2006, p. 17–18) defines three views of human knowledge that help frame academic inquiry. Each perspective describes the methods used to generate knowledge in relation to three phenomena: the natural world, human experience, and the artificial world. Inside the academy each perspective involves rigorous training in the appropriate methods of inquiry and an initiation into specific belief systems and values (Table 4.2).

Recognizing and honoring these perspectives, and having the agility to move between them, prevents faculty members, students, and their collaborators from becoming trapped in a single logic of inquiry.

The Integrative Power of Transdisciplinary Work

We have found that human-centered transdisciplinary methods can have several positive effects. First, by nature the subjects they address are often so complex that multiple inquiry methods are required to frame them. Participants tend to focus on defining problems and developing solutions rather than asserting the singular virtues of their disciplines. Transdisciplinary projects also become powerful forums for learning about disparate disciplinary methods, because participants see them in action as they are applied in concrete contexts.

Table 4.2 After Cross (Nigel Cross, 2006)

	In sciences	In humanities	In design
Phenomenon of study	The natural world	Human experience	The artificial world
Inquiry methods	Controlled experiment, classification, analysis	Analogy, metaphor, evaluation	Modeling, pattern-formulation, synthesis
Values	Objectivity, rationality, neutrality, and a concern for 'truth'	Subjectivity, imagination, commitment, and a concern of 'justice'	Practicality, ingenuity, empathy, and a concern of 'appropriateness'

Table 4.3 Characteristics of human-centered transdisciplines (Hillen & Levy, 2013)

Purpose	Improve the human condition in specific, concrete ways
Objective	Human-centered value creation (economic, social, cultural, etc.)
Foci	Defining problems, developing solutions, implementing
Stance	Proactive and action oriented, generating affordances
Primary competency emphasized	Know-how (integrating knowing and doing)
Content focus	Projects affecting the human condition.
Pedagogical format	Fieldwork, coaching and mentoring, staging learning activities in multiple learning environments, just-in-time knowledge acquisition.
Pedagogical orientations	Inspiration through empathy, ideation, action
Pedagogical tools	Observational research (such as ethnography), opportunity-finding, storytelling, strategic foresight, prototypes, user feedback, "reduction to practice."
Epistemology	Pragmatism and constructivism
Disciplines: students	Multidisciplinary
Disciplines: faculty	Multidisciplinary
Project definition	Context-based brief developed faculty, partners, and students. Subject refined or redefined as project proceeds.
Problem nature	Wicked: complex, ambiguous, messy
Mode of thinking	Collective and constructive
Outcome	Actions directed toward implementing appropriate solutions for problems in specific contexts.
Process type	Cyclical and iterative process
Evaluation	Self reflection, appropriate assessment methods to measure the particular outcomes in the real world
Location	In the field and staged project space

Characteristics of Human-Centered Transdisciplines
(Table **4.3**)

Innovation Literacy 4: The Agility of Learning and Acting by Integrating Knowledge with Things

Innovation leaders understand the central roles prototypes play as knowledge integrators. Think of prototyping artifacts as knowledge objects (Tsoukas, 2009, p. 169) that integrate knowledge, accrue meanings, and embody learning as they are iterated over time. Knowledge objects are powerful tools for productively integrating inductive, deductive, and abductive modes of reasoning.

> Prototypes encourage inductive thinking because they are concrete, situational and particular. The innovation leader and his or her collaborators can use these traits to reason from particular embodiments of an idea to general

(continued)

bodies of knowledge. For example, during the transformation of Philadelphia University into an innovation-driven institution the provost and his team had to consider how to realign the school's academic units to drive innovation by generating greater levels of collaboration, flexibility, and academic integration. Transformation at this scale is truly a wicked problem: complex, systemic, and rife with ambiguities. This redesign of the university was organized into a series of manageable and scalable prototyping activities starting with community charrettes and workshops that led to a prototype college (the transdisciplinary College of Design Engineering and Commerce), and eventually to restructuring the entire university into a system of three colleges integrated within a matrix. Throughout the process the Provost's Office worked extensively with the university community to continually and constantly map and model academic functions in novel arrangements until promising solutions could be tested with prototypes.

Prototypes also encourage general bodies of knowledge to be deductively applied to address the specific issues they represent. Using Philadelphia University again as an example, the transdisciplinary core curriculum (see Chap. 5) for the College of Design Engineering and Commerce began as a series of prototyping activities including trial courses, charrettes, workshops, modeling a new building to house the college, etc. These activities generated a core group of principles that were later applied to developing the curriculum and its assessment protocols.

Finally, prototypes are always in a state of iterative flux. This provisional nature conceptually softens their literal and representative qualities, allowing them to suggest novel possibilities for further development through abductive logic. Abductive reasoning is a creative process of explaining evidence by forming the most plausible hypotheses about it. These hypotheses can then be evaluated for validity. The power of knowledge objects is that they can directly interact with people and environments for which they were intended. Those involved in creating and developing them to solve specific problems can use these interactions as concrete evidence for forming abductive hypotheses about their success. The hypotheses can then be applied to further develop the prototype in the context of its actual use. If the process is well designed these cycles of iteration gradually lead toward the best solution.

Prototypes do not have to be objects. Academic program development itself can be framed as a prototyping process. Several faculty members at OCAD University and external collaborators formed a hypothesis based on their professional work, research, and interactions with students and other stakeholders that there was demand for a cross-disciplinary graduate program that focused on strategic foresight and innovation. The faculty members

(continued)

created an initial model for the program, including its mission and preliminary academic structure. In an effort to demonstrate the demand for the proposed model and receive user input for its development, 22 students were recruited, solicited for feedback, and preselected for the program well before its approval. The strong student interest strengthened the hypothesis that pent up demand existed for what was eventually named the Strategic Foresight & Innovation (SFI) masters program. The initial model was prototyped and tested by the first class of students after the launch year. Systems of formal and informal feedback allowed students and faculty to assess the prototype and iterate improvements. This assessment and iteration process continues today.

The agilities of collaboration, function, thought, and learning with things allow the innovation leader to find and solve problems by understanding the world from the points of view of those who inhabit it. Understood together, the agilities give the innovation leader the gift of strategic dissociation—a frame of mind that is radically open to seeing the world anew. In psychology, dissociation can be a pathology causing the afflicted to feel detached from the world. Yet to *strategically dissociate* is to willfully reframe the world and the issues faced by its inhabitants in order to empathize with them and look past our preconceptions of the problems they face.

Strategic dissociation can allow us to find meaning in ideas or methods that on the surface seems completely oppositional. Roger Martin (2009) defines this kind of integrative thinking as:

> The ability to face constructively the tension of opposing ideas and, instead of choosing one at the expense of the other, generate a creative resolution of the tension in the form of a new idea that contains elements of the opposing ideas but is superior to each.

Those who are concerned about the future of our universities must think past the binaries that are so deeply and structurally embedded in them. The list of these oppositions is extensive: generalist/specialist, applied knowledge/theoretical, knowledge, teaching/research, discipline/transdiscipline, qualitative/quantitative, artistic/scientific, and the list goes on. In order create universities for educating innovation leaders, all of us must reframe the tension in each binary as a generative force for "a new idea that contains elements of the opposing ideas but is superior to each" (Martin, 2009).

4.6 Changing University Ecosystems

Platform learning and the values and agilities it supports require new kinds of university ecosystems. What is an ecosystem? When we remove the normative interpretive framework for a traditional university, what Irving Goffman (1974,

p. 24) referred to as the "framework of frameworks," it becomes an overwhelmingly complex place. There are curricular and research agendas often spawning growth in solitary petri dishes, overlapping and sometimes ambiguous governance structures, hundreds of budget cost centers, diverse revenues, deep tacit cultural understandings that silently influence behavior, auxiliary support services, etc. Yet linking all of these actors and entities together are resource and knowledge flows, cultures of knowledge building and exchange, learning processes, and relationships with organizational entities beyond the university itself. Together they constitute a university ecosystem.

There are many interesting experiments occurring around the world that are building bodies of knowledge about how to design new kinds of university ecosystems that support platform learning. We will see in the next chapter how courses and programs at a variety of institutions are pursuing innovative pedagogies, challenging the hegemony of disciplinary thinking, inventing new ways of assessing students, reconceptualizing the spaces of learning, and bridging knowing and acting. While these course and program experiments provide important vantage points for reimagining universities that support platform learning, *we argue that university ecosystems must broadly change* in order to prepare innovation leaders to face the staggering complexity of systemic contemporary problems in disparate domains concerning issues such as global economic dynamics, climate change, healthcare delivery, and post-national policy development and conflict mitigation.

Scaling up

One approach to framing the kinds of ecosystems universities need to create is to look at the micro-level experimentation in teaching and research that is occurring in courses and programs pursuing the goal of creating innovation leaders. When conceptually scaled to the level of the university these experiments map the characteristics of the ecosystems we seek.

Most of the courses and programs we describe in Chap. 5 require broad interactions with knowledge sources and organizational partners beyond the walls of the classroom and university (see diagram below). They emphasize co-creative activities involving diverse teams of students, faculty, and outside collaborators. Frequently the activities involve many disciplines and often contain dominant transdisciplinary themes that cohere the learning and research processes. Frequently, learning and research requires outside funding from partners in business, government, foundations, or philanthropic individuals (Fig. 4.2).

The learning and research activities we describe above sound very much like those driving aspects of Etzkowitz and Leydesdorff's (2000) *entrepreneurial university,* where "teaching and research are expanded from traditional lectures and individual professor-student relationships into an experiential education and group-research format (Etzkowitz, 2013, p. 487)." This work focuses primarily on the roles universities are increasingly playing in ecosystems for regional development (particularly in technology and science) in partnership with government and business.

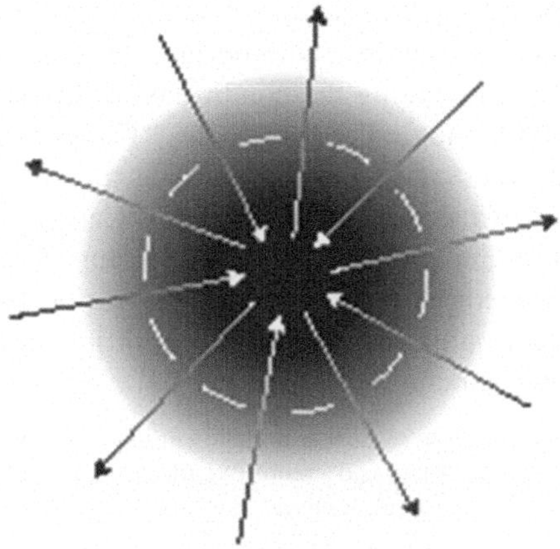

Fig. 4.2 Porous boundaries for fluid flows of knowledge

Etzkowitz and Leydesdorff (2000) discuss several characteristics of this *Triple Helix* innovation model that resonates with our description of teaching and research in Chap. 5. For example, academic activity in this model rejects the binary so typical of universities that rigidly separates applied and unapplied forms of teaching and research. Like many courses and programs focusing on creating innovation leaders, the Triple Helix innovation model seeks to incorporate multi-disciplinary research questions from the real world. The complex business and research relationships interweaving Stanford University and the surrounding Silicon Valley are an excellent of a Triple Helix ecosystem. The Triple Helix is a relatively new phenomenon and repositions the university from a producer of *basic* (unapplied) research that is then passed to industry for application to an active agent for innovation in the economy and society.

Multilateral Knowledge Creation
The proactive, engaged Triple Helix ecosystem model is consonant with our emphasis here on creating innovation leaders through teaching and research activities that emphasize situated action aligned with needs beyond the walls of the university. Our model rejects a linear progression of knowledge from the university to society and embraces a more interactive approach where multi- and transdisciplinary problems (research questions) from the real world inform teaching and research. While the Triple Helix model for university ecosystems is productive, its focus on innovation for economic development limits our use of it here.

Carayannis and Campbell (2011) posit an elaboration of the Triple Helix model by adding helices. The Quadruple Helix adds the contexts of civil society and the general circumstances of democracy conditioned by culture and the media. This additional context is important, because it recognizes the critical role of society at large in creating the conditions for knowledge production and innovation. On one level the logic here is simple, "The public becomes more integrated into advanced innovation systems. The public uses and applies knowledge, so public users are also part of the innovation system" (Carayannis and Campbell, 2011 p. 338). Yet this additional helix is also significant because it extends to a system of creative knowledge production and innovation by incorporating broader societal and cultural conditions and their supporting institutions that nurture creative knowledge environments:

> *knowledge of culture* and the *culture of knowledge,* values and lifestyles; multi-culturalism, multiculture, and creativity; media; arts and arts universities; and multi-level innovation systems (local, national, global)... (Carayannis and Campbell, 2011 p. 338, Fig. 4.3)

Finally, Carayannis and Campbell (2011) propose an additional, fifth, helix that gives context to the third and fourth helix models. The Quintuple Helix (see diagram) incorporates the natural environments of society and ecology. It allows us to frame the wicked problems of societal and environmental sustainability. The fifth helix encompasses the complex knowledge flows that occur between the nation state (government, civil society, culture) and academia as a means of framing a system of knowledge production and application—and innovation. These external

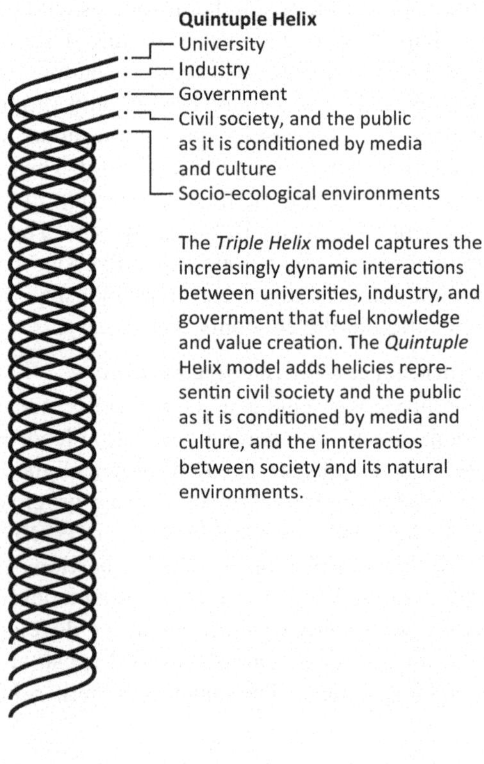

Fig. 4.3 Quintuple Helix

conditions of the Quintuple Helix ecosystem demand certain conditions of the internal university ecosystem:

> The Quintuple Helix is interdisciplinary and transdisciplinary at the same time: the complexity of the five-helix structure implies that a full analytical understanding of all helices requires the continuous involvement of the whole disciplinary spectrum, ranging from the natural sciences (because of the natural environment) to the social sciences and humanities (because of society, democracy and the economy) (Carayannis and Campbell, 2010 p. 62).

The logic of integrative innovation processes often characterized as variants of design thinking becomes understandable as both a cohering force in the expanding university ecosystem and a driver of its further development. The entrepreneurial university benefits from the two-way innovation process where knowledge and innovation move between academia and ecosystem components (government and industry). However, the potency of integrative innovation processes is particularly powerful when they confront the complexity of conditions described by the Quadruple and Quintuple Helix models.

These broader ecosystems provide us with the perspective to see how the innovation process derives energy and momentum by harnessing the cross-currents emanating from culture, social issues, environmental imperatives, education, economic conditions, government and politics, etc. The wider scope of the Quadruple and Quintuple Helix ecosystem models helps us understand the logic and necessity of multi- and trans-disciplinarity and the importance of mastering cross-sectorial collaboration that drives the production of knowledge for innovation. We have insisted in this chapter that integrative knowledge production processes (which include "making" centered pedagogies) require a deep understanding of and empathy with people as they live together in the real world and move through the human artifice. Carayannis and Campbell (2011) translate this imperative to the innovation ecosystem level when they describe multilateral flows of knowledge between the public (see p. 338), universities, and other ecosystem entities.

4.7 Hope

The higher education sector has in the past decade focused considerable time and treasure on small to mid-scale innovation that can be characterized as shifts in the experience of students and the platforms they use to learn and manage their education (Keeley, Pikkel, Quinn, & Waters, 2013). Such advances are very important, but much less effort has been invested into perhaps what existentially matters the most: remaking the very model of value creation that defines universities and drives them forward. Simultaneously innovating in the realms of the value creation model, student experience, and platforms could truly transform the higher education sector.

Universities should confront the reality that the value (economic, social, cultural, scientific, etc.) they were originally designed to create primarily served a class of problems confronted by an earlier industrial society—not ours. This is not to diminish the tremendous creative energy that is being invested in innovation-centered programs around the world, but these valuable initiatives often exist as grafts on tradition models of value creation. Such grafts take the form of academic programs with liminal positions in universities: special institutes and centers, "collaboratoriums," incubators, capstone experiences, sponsored projects, etc. A few important institution-level innovation initiatives have also been launched, including the ambitious national effort behind Aalto University in Finland, and the redesign of Philadelphia University in the United States.

Our goal should be to use these efforts for envisioning and implementing pervasive infrastructures for innovation that encompass entire university organizations, cultures, and their broader ecosystems. The momentum toward a new kind of university that broadly supports innovation leadership gives us tremendous hope. Let us learn from the innovation initiatives occurring around the globe, scale them when appropriate, and use our accrued know-how to redesign university value creation models and the experiences and platforms they support.

References

Arthur, M. B., Defillippi, R. J., & Lindsay, V. J. (2008). On being a knowledge worker. Organizational Dynamics, 37(4), 365–377. Retrieved from http://ezproxy.philau.edu:2048/login?url=http://search.ebscohost.com/login.aspx?direct=true&db=buh&AN=35725552&site=ehost-live&scope=site

Berggren, e. a. (Ed.). (2011). Knowledge integration and innovation. Oxford: Oxford University Press.

Carayannis E, Campbell D (2010) Mode 3 Knowledge Production in Quadruple Helix Innovation Systems, 21st Century Democracy, Innovation and Entrepreneurship for Development. Springer, NY

Carayannis, E. G., Campbell, D. F., & J. (2011). Open innovation diplomacy and a 21st century fractal research, education and innovation (FREIE) ecosystem: Building on the quadruple and quintuple helix innovation concepts and the "mode 3" knowledge production system. Journal of the Knowledge Economy, 2(3), 327–372. doi:http://ezproxy.philau.edu:2887/10.1007/s13132-011-0058-3

Cross, N. (2006). Designerly ways of knowing. London: Springer Verlag.

Darnton, R. (2008). The library in the new age. New York Review of Books, 55(10), August 18, 2014.

Davidson, C. (2011). Now you see it: How technology and brain science will transform the way we live, work, and learn. New York: Penguin Books.

Dewey, J. (1925). Experience and nature Open Court Publishing Company, NY

Etzkowitz, H. (2013). Anatomy of the entrepreneurial university. Social Science Information, (52), 486.

Etzkowitz, H., & Leydesdorff, L. (2000). The dynamics of innovation: From national systems and 'Mode 2' to a triple helix of university. Research Policy, 29(2), 109. Retrieved from http://ezproxy.philau.edu:2048/login?url=http://search.ebscohost.com/login.aspx?direct=true&db=buh&AN=2815434&site=ehost-live&scope=site

Goffman, E. (1974). Frame analysis. Lebanon, NH: Northeastern University Press.

Hargadon, A. B. (2002). Brokering knowledge: Linking learning and innovation. Research in Organizational Behavior, 24, 41–85. Retrieved from http://ezproxy.philau.edu:2048/login?url=http://search.ebscohost.com/login.aspx?direct=true&db=buh&AN=12342054&site=ehost-live&scope=site

Hargadon A, Sutton RI (1997) Technology brokering and innovation in a product development firm. Administrative Science Quarterly 42(4):716–749, Retrieved from http://ezproxy.philau.edu:2048/login?url=http://search.ebscohost.com/login.aspx?direct=true&db=buh&AN=61521&site=ehost-live&scope=site

Hillen, V., & Levy, P. (2013). People, place, process: Lessons learnt on the way to a d.school. International Conference on Engineering Design, Seoul.

Keeley L, Pikkel R, Quinn B, Waters H (2013) Ten types of innovation. John Wiley & Sons, Hoboken, NJ

Martin R (2009) The opposable mind. Harvard Business Press, Cambridge

Pine, B. J.,II, & Gilmore, J. H. (1998, Jul/Aug 1998). Welcome to the experience economy. Harvard Business Review, 76, 97–105. Retrieved from ABI/INFORM Complete, ProQuest

Research Library database. Retrieved from http://ezproxy.philau.edu:3873/docview/227777207?accountid=28402

Schön D (1983) The reflective practioner. Basic Books, New York

Siemens, G. (2006) Knowing knowledge, Creative Commons License.

Tsoukas H (2009) Creating organizational knowledge dialogically: An outline of a theory. In: Richards T, Runco M, Moger S (eds) The routledge companion to creativity. Routledge, New York, pp 160–176

Chapter 5
Charting Interdisciplinary Innovation Programs: Map of Experiences

Paola Bertola, Nabil Harfoush, and Federica Vacca

5.1 Introduction

There is consensus that higher education institutions in post-industrial economies are experiencing a general crisis, expressed in their lack of flexibility to address contemporary challenges and rapidly changing needs (as touched upon in Chaps. 1–3). A literature review shows that the debate on cross-, inter-, multi-, and transdisciplinary education is still raging among academics (Kozma, 2005). Within this debate a theme has emerged during the last decade for schools of art, design, engineering & applied sciences, centered around the subject of enabling innovation and positive world transformation by design.

Innovation enabling became its imperative and its most progressive institutions, such as Stanford's d.school, IIT in Chicago, School of Information at Berkeley, Media Lab at MIT (Dym et al. 2005, Litzinger et al. 2011), became laboratories for testing new pedagogies and educational approaches as well as supporting the transition from discipline-based education to process/problem-focused education (Beetham, and Sharpe 2013). As introduced in Chap. 4, the needed transformation should be radically centered on students and based on innovative ways of teaching, new organizational frameworks and collaborative learning environments.

During the last decade, many examples of these changes emerged from all over the world, supported by a scientific debate on how to positively drive different national educational systems to be able to face the fast and turbulent transformation of economies and societies (Anderson 2012; Deeun and Sunhee, 2013). One of the most evident and earliest changes was related to universities opening new programs

P. Bertola (✉) • F. Vacca
Design Dept. Politecnico di Milano, via Durando 38/a, 20158 Milano, Italy
e-mail: paola.bertola@polimi.it

N. Harfoush
Stategic Foresight and Innovation Masters Program, OCAD University, 10 McCaul St., Toronto, ON, Canada, M57 1W11

B. Banerjee, S. Ceri (eds.), *Creating Innovation Leaders*,
Understanding Innovation, DOI 10.1007/978-3-319-20520-5_5

not anymore centered on creating experts in specific domains of knowledge but rather centered on problems. In this perspective many curricula addressing complex topics such as system design, service design, sustainability or even innovation itself were created, such as for example the programs introduced at Glasgow University of the Art in Scotland on "Design Innovation & Environmental Design", "Design Innovation & Service Design" and "Design Innovation & Citizenship". In these new programs the goal of education is enabling individuals to be responsible for their future role in society and for its positive transformation, such as in the Graduate School for Future Strategy created at KAIST University in Seoul, Korea, which claims:

> With a basis in new technology and knowledge fusion, our country must switch to a paradigm leading future society. To accomplish this goal, we must foster creative talent to lead us through uncertainties with long-term strategies for at least 20 years [. . .] . "[1] It strives to deliver a top global program able to solve global and humanitarian problems with an insightful perspective.

In many cases not just programs but new schools and departments were created within existing universities as a means for giving **porosity** to the system. This was achieved both by integrating different disciplines previously belonging to separate entities, and by connecting institutions, that are able to provide a rich, multicultural experience to students enrolled. If the d.school in Stanford represented one of the first experiments breaking boundaries in the traditional model of schools and programs based on exclusive discipline, many other cases followed, such as the i. school at Tokyo University,[2] built on that example, or others looking even outside their own institution to bridge different universities (Ito et al. 2014; Kurokawa, 2013). This is the case, among others, of the "EIT ICT Labs Master School" involving 20 partners in Europe, and "The Global Leadership Programme" at the Singapore University of Technology and Design,[3] which involves also MIT Boston as a partner (Dimmock and Goh 2011).

Breaking disciplinary boundaries and facilitating **knowledge integration** drove many academies to reform their traditional departmental structure, as was the case for Aalto University's School of Arts, Design & Architecture in Helsinki, Finland, or led to the creation of a new entity entirely focused on multidisciplinary education such as the Interaction Design Institute in Copenhagen, Denmark.[4] In this global scenario an increasing value has been given to teaching practices and their transformation. This led to experimenting with different pedagogies centered on **learning in the field** and project-based education, as well as collaborative learning and

[1] http://www.kaist.edu/html/en/edu/edu_030106.html

[2] http://ischool.t.u-tokyo.ac.jp/english: "We cultivate to educate a new style of leadership. This encourages leaders to serve others while staying focused on achieving results in line with a broad view to business and society, i.e. they conceive innovative ideas, collaborate with a variety of stakeholders, and facilitate to realize changes."

[3] http://www.sutd.edu.sg/glp_2014.aspx

[4] http://ciid.dk/

teaching, such as in the School of Design Thinking at the Hasso Plattner Institute, Potsdam University, Germany,[5] or in the Creative Intelligence and Innovation School at the University of Technology in Sydney, Australia,[6] or, finally, at the University of California in the School of Information (I-School), the youngest and smallest school in Berkeley that states:

> I School is a graduate research and education community committed to expanding access to information and to improving its usability, reliability, and credibility while preserving security and privacy. This requires the insights of scholars from diverse fields — information and computer science, design, social sciences, management, law, and policy. [7]

Different teaching approaches focused on the active role of students, exposure to real problems and multidisciplinary working teams were also supported by the tentative of incorporating more flexibility in university curriculum, where sometimes these novel innovation programs are often offered as complementary or in addition to students' regular paths, such as Alta Scuola Politecnica in Milan and Turin, Italy, Stanford's d.School, École des Ponts, Paris-Est University, France, or in the "Cologne Model for Design Education" implemented at Köln International School of Design, where:

> Projects replace traditional »classes« [...] and students have the possibility to compose their own curricular program, by choosing long, medium, or short term projects and seminars.[8]

As underlined, it is possible to detect in the scenario described above, many of the requirements identified in Chap. 4 for transforming universities and enabling them to educate innovation leaders. Therefore, the present configuration of universities, their experimental bottom-up approach, and their diverse cultural and scientific heritage show the complexity of an incomplete and ongoing phenomenon.

5.2 The Como Innovation Summit: Taxonomy of Experiences

A number of higher education organizations, which are transitioning towards the novel paradigms introduced in Chap. 4, created the Como Innovation Summit initiative as a platform for sharing their early experience and lessons learned. The two summits were not just attracting academic participation: in addition, there were a number of industry partners and sponsors attending. However, education was a common concern of most of the participants, and therefore the focus of most discussions.

---

[5] http://www.hpi.uni-potsdam.de/d_school/home.html

[6] https://www.uts.edu.au/future-students/creative-intelligence-and-innovation

[7] http://www.ischool.berkeley.edu/courses

[8] http://kisd.de/en/kisd/history/

One of the activities of the Como Summit group was to survey the participant academic organizations in order to recognize any patterns in this emergent phenomenon, with the aim of codifying possible new paradigms for education becoming itself a central and responsible actor in the positive transformation of contemporary society. The survey is therefore qualitative research aiming at gaining insights into the group of participating institutions rather than a statistically significant study on any regional scale. The data obtained allowed us to map the innovative academic programs in many dimensions: by geography, type of program, delivery model, content, diversity of student body, size of faculty etc.

One additional objective of the survey was to verify whether the principles and characteristics of the necessary trans-disciplinary education developed in the preceding Chap. 4 are actually observed in the field and to what extent. The following sections present major findings from that survey, organized by various dimensions explored.

5.2.1 Como Summit Geography and Timeline

As Figure 5.1 shows, the majority of participants were from western higher education institutions (European and North American), with one Asian institution in attendance.

The composition of the participants reflects the natural networks of the founders of the Como Innovation Summit. The institutions represented are mainly from the higher education sector with a notable clustering of art & design schools together with engineering and science schools. The absence of educational institutions from the medical and biological sectors may have its roots in the fact that the medical sector, faced with the highly complex human body, had started already in the 1970s

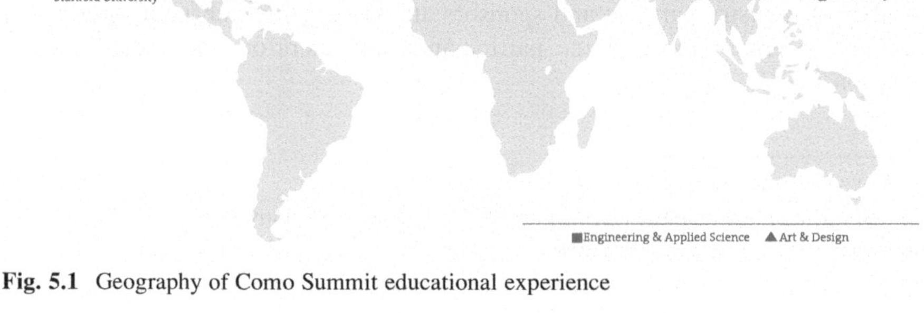

Fig. 5.1 Geography of Como Summit educational experience

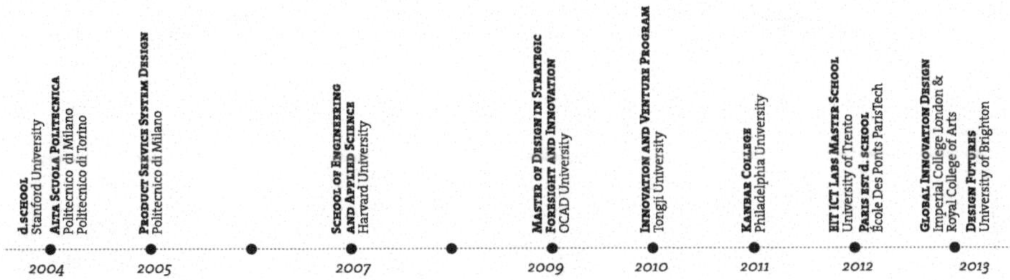

Fig. 5.2 Timeline of programs

discussing multi-disciplinary education and collaboration. It could be hence argued that that sector was ahead in that process and therefore not interested in joining a similar but emerging process in other sectors. During the two summit meetings, several participants revealed advanced plans to launch trans-disciplinary programs in healthcare-related domains.

Overall, the data collected encouraged the hypothesis that the trans-disciplinary education paradigm is globally emerging as an important phenomenon. It seems to be increasingly addressed as a goal for transformation by all participants, with the majority of programs having been established in the last five years as shown by Figure 5.2.

More information was collected by means of an extensive survey. The following sections analyze the educational models captured through the participant survey across a number of dimensions, to show evidence of the evolving picture of educational transformation.

5.2.2 *Survey Outcomes*

5.2.2.1 Types of Programs: Multilevel and Flexible Structure

As Figure 5.3 shows, the content offered in programs seems to be spread across the entire spectrum of postsecondary education. The course content was mostly at graduate level (Master and PhD level) with roughly a third at undergraduate level.

Interestingly enough, this content was delivered through a wide range of programs, from certificate and diploma to bachelor and master degree programs, with 60 % being offered in bachelor and masters programs and most of the rest in certificate and graduate diploma programs.

The concentration of the content at graduate level indicates that in the current paradigm a foundation in a particular discipline is perceived to be a prerequisite for building upon the transdisciplinary knowledge. The fact that in the relatively small group surveyed at least one institution is offering a multidisciplinary program at the doctoral level indicates that this emerging domain is starting slowly to penetrate mainstream education.

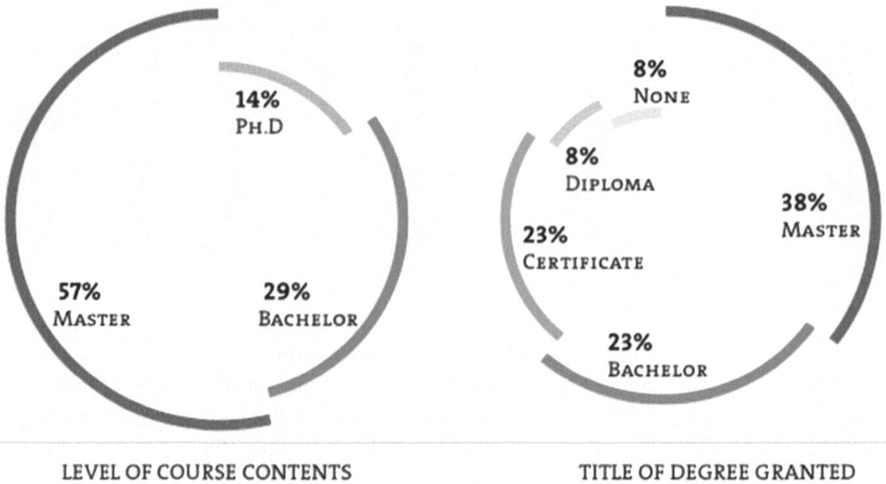

Fig. 5.3 Level of course contents and title of degree granted

Some institutions offered such content independently of any degree program (d. school, for example), making the content accessible to a wider range of students from within and outside the academic institutions. In many cases, courses/programs offered led to the acquisition of simple certificates or even no certification (8 %) at the end by students (Figure 5.3). This points at the cross-disciplinary orientation of these types of educational activities, but also at their relative impact on the mainstream educational structure of each institution (Figure 5.4).

The duration of the various programs that offered a degree upon completion was generally the standard three to four years for bachelor programs and two years for masters programs. Programs that did not offer any degrees, such as the Stanford d. school or the d.school at Paris-Est University, had a very wide spread of durations, from one-week intensive courses to full-fledged programs over 36 weeks, which reflects the greater accessibility and flexibility these programs are aiming to achieve (Fig. 5.5).

As shown by Figure 5.6 almost all of the surveyed programs (95 %) are offered in English even though only 50 % of the institutions are from English-speaking countries. This confirms the general trend of English as a universal communication language in trade, science, technology and interpersonal communication globally.

5.2.2.2 Student Communities: A Highly Diverse Collectivity

As shown in Figure 5.7 all surveyed programs had significant non-local student population confirming the high demand existing globally for this type of education and supporting the hypothesis of a global emergent phenomenon. The overall results for non-local students population confirms that the global trends of Asian

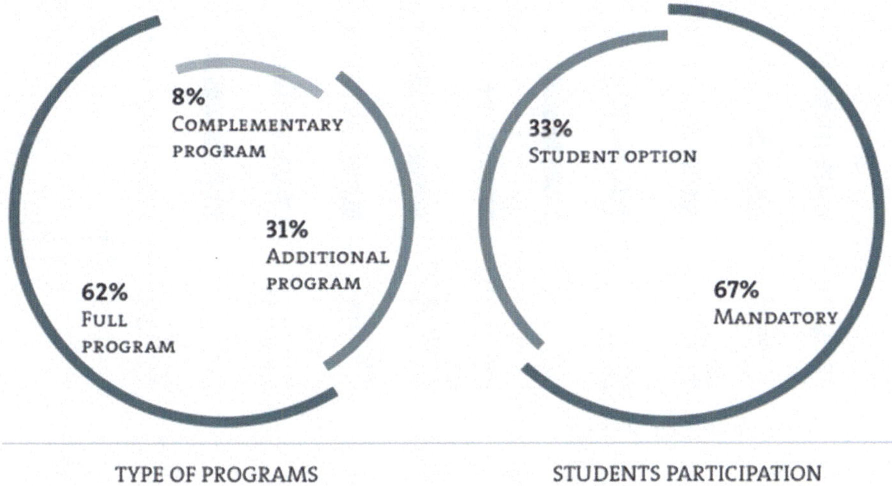

Fig. 5.4 Type of programs and modalities for students participation

Fig. 5.5 Duration of programs

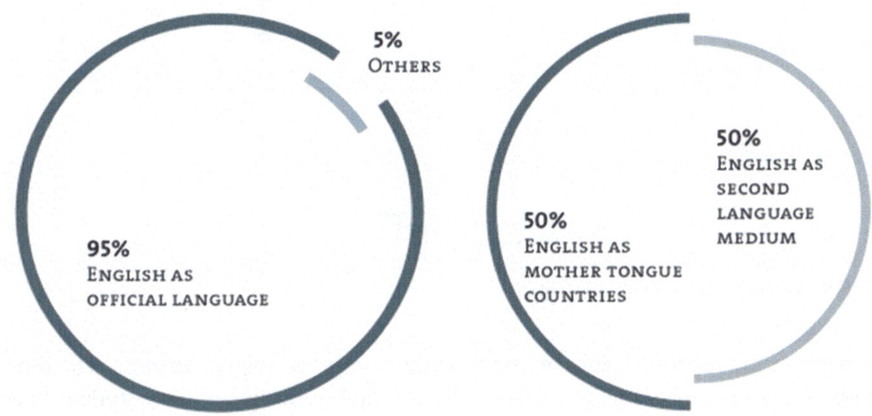

Fig. 5.6 Official program language

students (42 %) seeking the experience of studying abroad, is not affected by the uncommon and innovative approach of the surveyed institutions (Figure 5.8).

As illustrated by Figure 5.9, the recruitment of students in the innovative programs surveyed fell into two broad categories. One group of programs was focused on offering multidisciplinary curricula to students in specific vertical disciplines,

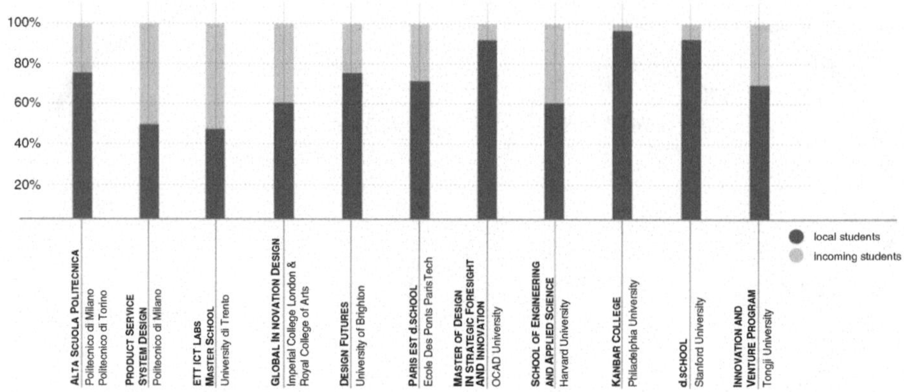

Fig. 5.7 Local and foreign students

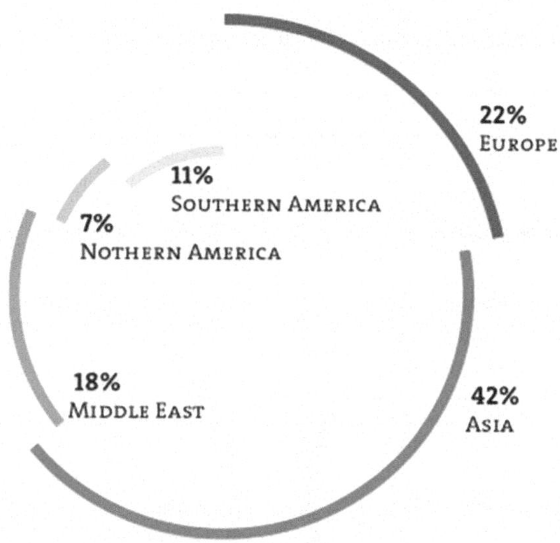

Fig. 5.8 International student participation

for example students in design or architecture or engineering or business. The other group offered its programs to a highly diverse and multidisciplinary student body recruited from many different disciplines.

Those programs that were not bound to a specific degree but were positioned as electives, complements, or additions to other academic programs were able to recruit from a much broader population (continuing education candidates, lifelong learners, executive trainees, and professional accreditation or maintenance candidates, etc.) while maintaining stringent quality criteria.

At least one program (Strategic Foresight & Innovation Masters at OCAD University, Toronto) has institutionalized the diversification of its students through

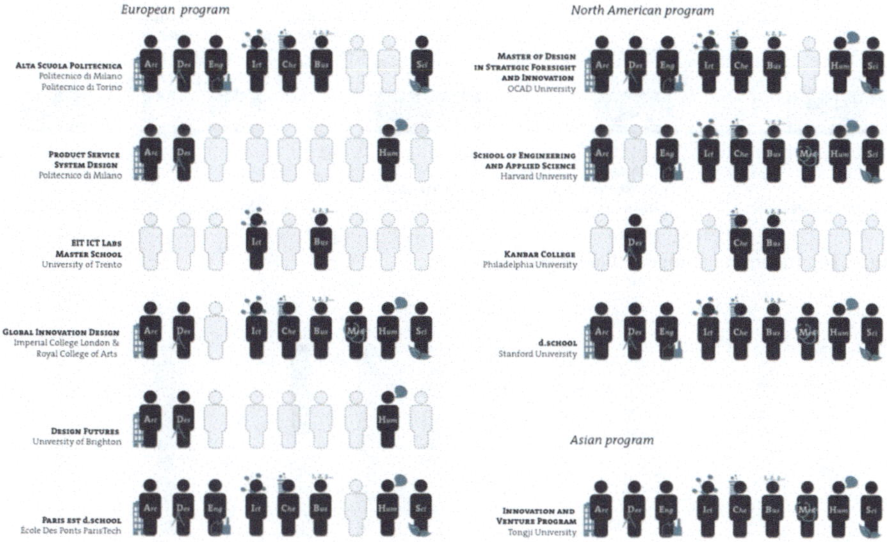

Fig. 5.9 Diversity of students' disciplinary backgrounds

an approach called Designed Diversity, which evaluates candidates not only on their academic performance but also on their contribution to the diversity of their class or cohort in any dimension. In conversations with various participants of the summits, the higher diversity in the students of this group was perceived as one important element in moving from a multidisciplinary to a transdisciplinary mode in program delivery.

The most frequent disciplinary background found in the students of the second group of programs is "design" followed closely by architecture, business (including economics and law) and humanities. Engineering, technology and science are the next frequent backgrounds (Figure 5.10).

A notable absence is the lack of participation by students with a medical background. It is not clear if this is caused by normal biases in the process of inviting institutions to the summit, or by the different structure of medical education. Chap. 4 touched upon the emergence of multi-disciplinary thinking and debate in the medical education much earlier (in the 1970s), which could also be an explanation for this absence. Whatever the case may be, there is much merit in targeting students and faculty from the medical and biological sectors in order to leverage their earlier experience in this domain and expand the diversity range at the same time.

Despite the encouraging emergence of many multidisciplinary programs, the number of students enrolled in such programs constitutes still in most institutions a very small fraction of the total enrollment. In large universities the percentage of enrolled students in such innovative programs with regard to their total enrolment was in the range of only 0.2 - 0.6 %. In smaller universities that percentage was in

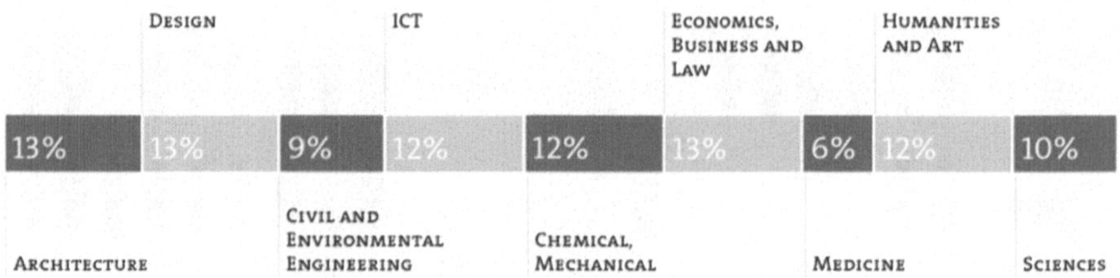

Fig. 5.10 Incoming students' backgrounds

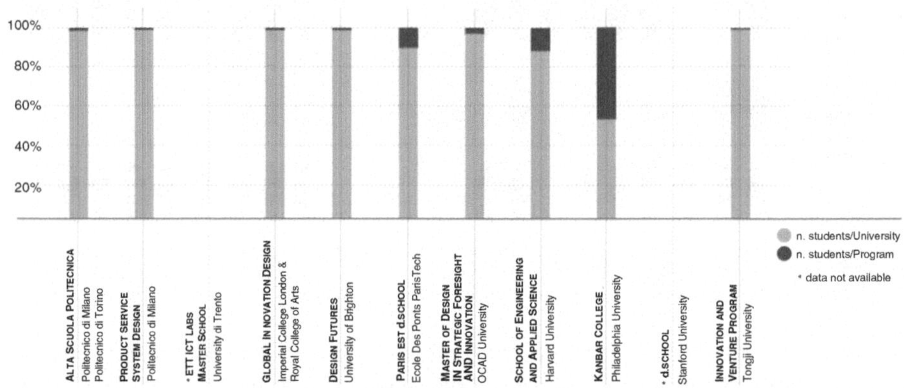

Fig. 5.11 Impact of innovative programs on their institution

the range of 10 to 15 %. These numbers illustrate the long way still to go for the new models of multidisciplinary education to reach the majority of the student body and have a significant impact on human society (Figure 5.11).

5.2.2.3 Faculty Staff: A Hybrid Community

A common characteristic of all surveyed programs is their use of external and diversified faculty to support the delivery of their content. The measure of participation of such external resources and their nature varies widely even within the small group surveyed. While the average of external instructors is 40 % (Figure 5.12), there are institutions that used almost exclusively internal faculty supported by limited invited guest speakers, and also programs that used up to 80 % of industry or external resources in their delivery. The nature of these external resources also varied considerably: industry ambassadors, guest lecturers, visiting professors, contract instructors, adjunct faculty, appointed part-time faculty with their own external professional practices, etc. The broader the multidisciplinary structure of a program, the more the difficulty an institution will face providing the knowledge and expertise for such broad delivery from within the institution. The wide range of external resources is a response to this challenge and a creative

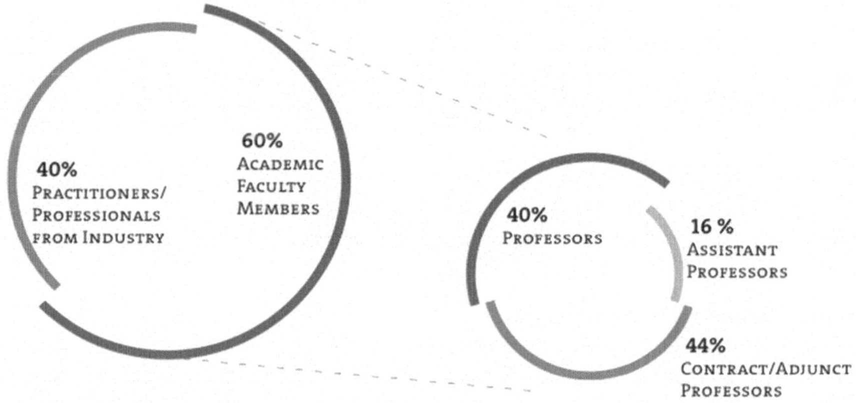

Fig. 5.12 Average composition of program faculty

workaround to bypass the various administrative and financial barriers resulting from the old academic structure that inhibit adapting to new needs. This creativity is also contributing to increasing the porosity of the barriers between the institutions and the outside world.

5.2.2.4 Pedagogical Focus: The Training by Doing Model

A relevant part of surveys was focused on codifying teaching methodologies, detecting possible trends, and gaining insights in the nature of classes offered by innovative programs. The survey probed the main educational methodologies applied in the analysed programs across two dimensions: use of project-based learning and balance between theoretical and applied contents in each course/activity offered.

As pointed out by Figure 5.13 all programs showed a bias towards project-based learning, which is consistent across the various geographies of the surveyed institutions (Europe, North America and Asia). Moreover, an analysis of each course, studio, and/or activity offered in the surveyed programs shows a consistent predominance of project-based units delivered under theoretical contents, with some cases being totally based on a field project model, such as in the Global Innovation Design joint programmes of Imperial College and Royal College of the Art, London (Figure 5.14).

In terms of pedagogical focus the survey mapped the relevance of different outcomes required by project-based activities, including problem setting, idea generation, execution and implementation, and marketing and exploitation, in a scale going from collateral to core. The result of this mapping, reinforced by the free text explanations obtained through the survey form, underlined that at the core of the new type of programs are the early stages of the design process (Figure 5.15). In fact, in almost all programs a strong commitment is expected from students to reframe the problem given at the initial brief, with the aim of uncovering more

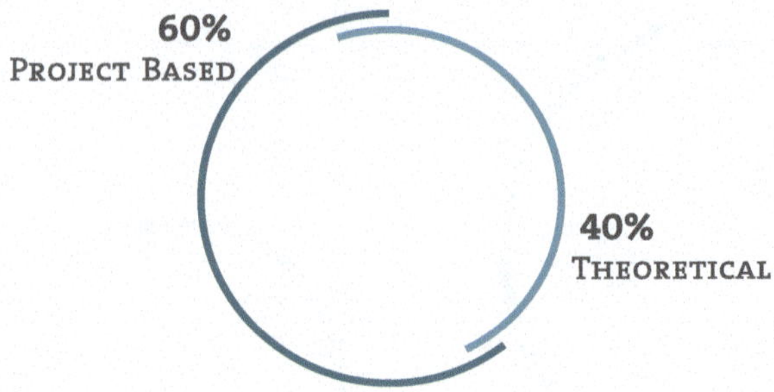

Fig. 5.13 Theoretical vs. project-based

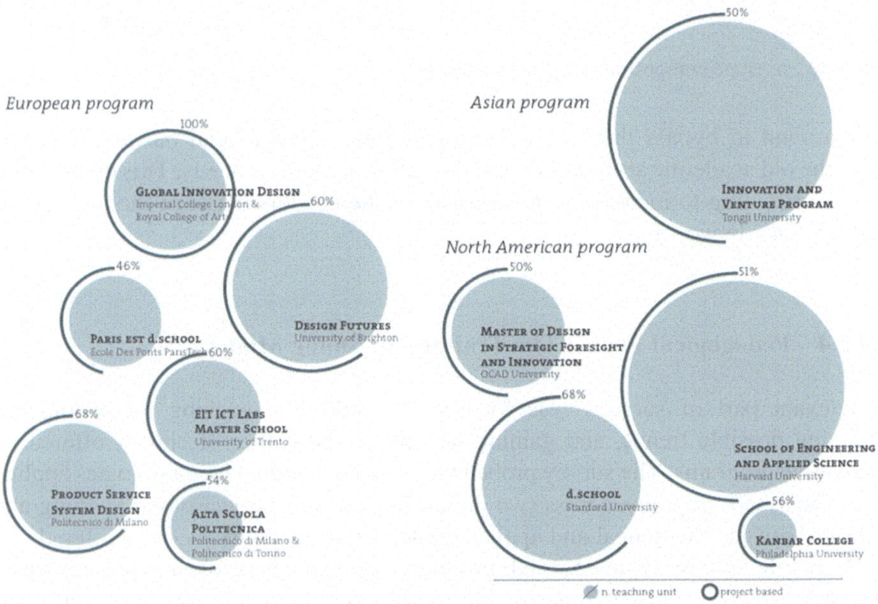

Fig. 5.14 Theoretical vs. project-based teaching units delivered

important and fundamental problems before generating new ideas. While execution and implementation of concepts remain very relevant, they don't seem to be a core requirement. Finally, less importance is generally given to marketing and exploitation, probably because of the highly experimental and exploratory nature of topics and projects chosen (Figure 5.16).

5.2.2.5 Pedagogical Approaches: From Problem-Centered to Process-Centered

The surveys also detected the main subjects addressed by courses and projects within the different programs (Figure 5.17). This specific analysis reinforces the

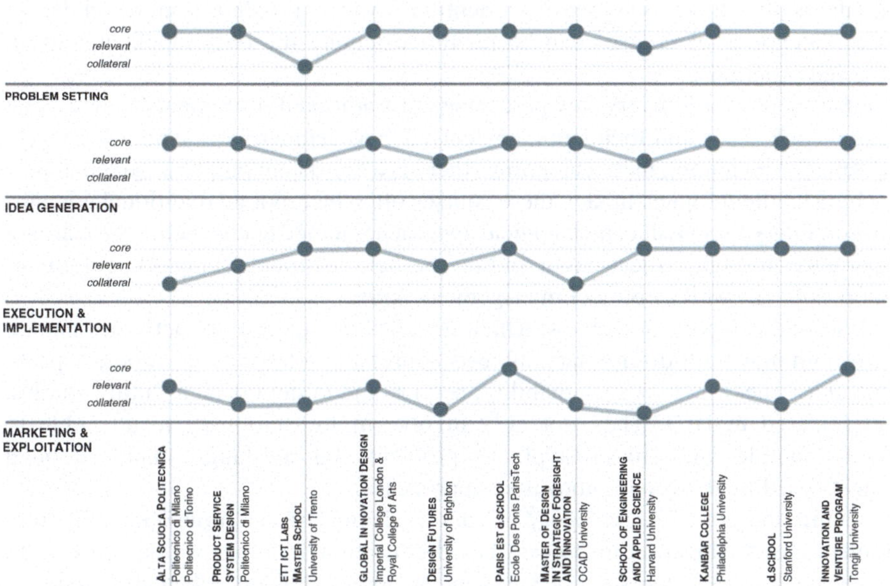

Fig. 5.15 Pedagogical approaches

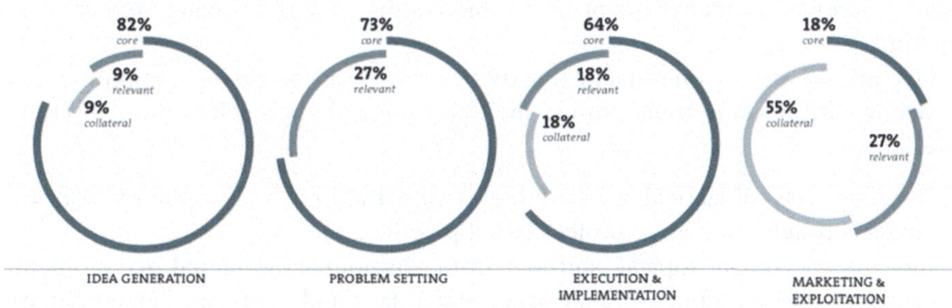

Fig. 5.16 Educational goals

Fig. 5.17 Emerging topics in surveyed programs

hypothesis that there is increasing attention towards problem-centered didactic. Three categories of courses can be identified among the ones taught within all surveyed programs.

First, there is still a very low percentage of traditional disciplinary courses that often include basic and foundation content, such as geometry or computer science. However, even among these categories, there is a consistent emphasis given to new teaching methodologies. In fact, these courses often incorporate traditional domains of disciplinary knowledge, pedagogical approaches aimed at enhancing the learning experience through project-based and on-the-field learning, teamwork, communication skills, as well as project management skills.

A second category of courses, which includes the majority of activities taught, focuses on new multidisciplinary subjects addressing relevant contemporary problems and issues, such as sustainability, service and system design, future studies, user-centered & HCI. They not only involve multidisciplinary faculty able to address complex and cross-disciplinary problems, but are taught mainly through project based methods and innovative approaches.

A third and not trivial category of courses demonstrates an ongoing shift from problem-based education to process-based education. In other words, there is an increasing focus of courses not anymore on topics, but on enhancing students' aptitude for putting in place innovative cognitive processes while facing any type of problem. This is the case, for example, in several courses focused on decision-making, innovation management, social innovation, design thinking, and creative thinking.

Overall, the survey on didactical activities shows an increasing commitment of academic institution to transforming their pedagogical approach at three different levels:

- Bringing new didactical methodologies into traditional disciplinary courses, taught through innovative project-based practice.
- Innovating course topics, shifting from disciplinary-centered to problem-centered classes/projects, addressing emerging and complex contemporary issues.
- Innovating the entire pedagogical goal, aiming at training students for becoming innovation leaders, focusing not anymore on topics/problems but on cognitive processes.

These three levels of pedagogical transformation aren't exclusive in each institution, but often mixed into educational models that can be slightly different from one another. Many of these differences have a deep connection with the contexts and constrains faced during the programs creation. However, there seems to be a strict relationship between the type of student community approached by the program and its pedagogical focus. The third level is more represented by institutions, such as Alta Scuola Politecnica, Stanford's and Paris' d.schools, or Toronto's SFI program that were able to cross the boundaries of disciplinary schools, departments and even institutions, involving and attracting students from very different contexts. Having this multidisciplinary student body they were compelled to provide educational experiences enhancing students' cognitive attitudes in being agents of innovation in any given context, more than providing them with specific bodies of knowledge, even if integrated and belonging to different disciplines.

5.3 Challenges and Open Issues

The survey data along with the interviews and discussions with participants of the two Como Summits has focused our collective understanding of the current state of the art, including challenges and barriers faced. This section examines the common patterns in the current model of higher education in the hope of supporting a more effective and coordinated approach towards removing the barriers and focusing leaders' attention on the critical issues that must be addressed if transformative changes are to be achieved.

We have chosen to group the challenges and open issues under four subtitles, although it should be clear that many if not all of these issues are intertwined and interdependent.

1. **Impact and Content Issues**
   One observation across most institutions is that the innovative programs are still a tiny portion of the overall education programs offered by each institution; this is true in the number both of enrolled students and of faculty involved in these programs. This low penetration rate is therefore not having enough impact on the rest of the institution or on the external ecosystem. This is not to say that the programs do not have an impact at all but rather that the impact is still insufficient to achieve the kind of changes required to address pressing social, economical and environmental issues of our time.

   In many instances the programs are seen by the rest of their institutions, including the administration, as experiments and explorations in future pedagogy that are remote from wider implementation. In other instances they were viewed as a useful tool to help brand the entire institution as different and progressive. Interestingly enough, even when a design school was involved in a multidisciplinary collaboration, not much effort went into the design of the collaboration itself and its embedded processes, which led to falling back onto traditional siloed processes and clashing of the different frames and methodologies.

   The evolution towards new models of pedagogy is being initiated from the bottom up, often without significant support from the administration and at times even with some resistance. Several started as innovative course proposals and expanded gradually to form a relevant part of a program. The lack of executive buy-in and support is leading to long development and deployment times relative to the few instances where executive support was fully available and driving the transformation. This underlines the importance of presenting convincing cases to institutional leaders to convince them of the imperative for change, and offer a path with mitigated risks to the transformation. Gaining executive buy-in and support is a major accelerator of the emergence of new pedagogy and the transformation of the system.

Several institutions attempted to explore the multidisciplinary education space by simply bringing together educators from a number of disciplines, but in the absence of a solid unifying framework to provide the "glue" between the various disciplines, these attempts did not deliver the results hoped for and at times created a negative backlash against the multidisciplinary approach. Design thinking is one of those unifying frames that might help converge the various disciplinary participants towards a cohesive program. Systems thinking is another such unifying frame that has presence and acceptance in many disciplines and therefore can help such convergence. There is consensus that any convergence effort must start by systematically and patiently establishing a common language and vocabulary among the participating disciplines.

Another issue faced by multidisciplinary programs is the perception by each "vertical" discipline that these programs lack the rigor as defined and practiced in each discipline. There is an urgent need to establish multidisciplinarity as a "horizontal" discipline in itself, with its own definitions and practices of rigor that are no less valid than any of the vertical disciplinary approaches to rigor.

2. **Capacity and Resource Issues**
   Many of the multidisciplinary initiatives are started bottom-up by few individuals, often within a resistive or reluctant environment. Consequently, the individuals and teams taking the initiative face substantial difficulties in terms of capacity and available resources to develop the necessary ecosystem required for innovation both internally within the organization and externally through alliances and partnerships (see Chap. 2).

   Another challenge for the innovators initiating projects of multidisciplinary courses and programs is to find the required teaching competences. Often the broad diversity of disciplines aspired to cannot easily by found within the walls of the institution, but hiring resources from outside the institution is usually a slow and cumbersome process. Compounding this challenge is the need for multidisciplinary instructors to bring significant soft skills in addition to their disciplinary hard skills. Many of these innovators efforts falter against administrative and regulatory rules enacted in their organization. More of this aspect is discussed under the next section.
3. **Administrative and Organizational Issues**
   The transformation efforts in the domain of new pedagogy and multidisciplinary education have been focused mostly within the academic structures of the institutions, but there was consensus that the new programs required an array of support in order to succeed, including: agile processes; new ICT systems; different admission policies; different hiring and advancement rules; and often more flexible financial processes. All these requirement fall usually under the purview of the administrative structures of the institutions, which more frequently than not are not engaged and hence do not participate in the bottom-up transformation initiatives. Consequently, most of the above listed support is not

available to the initiatives emerging within the institution and as a result severe tensions arise between the people leading these initiatives and the administrative and governance functions of the institutions. To avoid such situations it is critically important to engage and involve administrative leaders early on in the process and to gain their support and sympathy.

It is as important to understand the constraints under which administrators have to operate, particularly financial constraints and limitations rooted in legal and regulatory aspects that administrators may have little control over. Several participants mentioned that once they engaged administrative leaders successfully, these leaders displayed creativity in finding "workaround" solutions to such regulatory or legislative barriers.

Another interesting set of learned lessons touches on the importance of managing change wisely. Even when the support of the institution's executive leadership is available, any organizational changes in process or structure face substantial pitfalls and barriers. One of the challenges is that creating a new structure does not automatically cancel or remove deeply embedded tacit assumptions that existed in the previous structure. Only a highly participative process of co-creation can elicit and address these assumptions.

Another barrier when introducing organizational changes is that faculty members and administrators alike have difficulties understanding their identity within the new structure. Again, understanding this new identity and proactively accepting it cannot be attained without a highly participative and inclusive process that allows reasonable time for this adaptation process to take hold (see the invited chapter of Philadelphia University for a good discussion of these aspects).

Interestingly, students face similar challenges understanding their new multidisciplinary professional identity in the context of traditional lines of business and corporate functions. Similarly, employers have difficulties labeling graduates of multidisciplinary programs in the context of their hiring practices. Several participants reported that only when employers observe such graduates in action do they drop their need to label a candidate and seek the new and superior skill set that the graduates display. Some institutions created structures that help demonstrate the new skill space of students and graduates to potential employers. Embedding students and graduates in the teams of real-world projects facilitates the recognition of these new skills by industry, government and the social sector (see the invited chapter by OCAD University's SFI program and the role of Strategic Innovation Lab in this respect).

4. **Cultural Issues**

As in every substantial change process the surrounding culture plays a critical role. For example, when trying to introduce design thinking to faculty in various regions, one participant found significant attitude differences between the US ("let's do it"), Canada ("who's done it before?") and Europe ("let's think about it"). Other participants observed resistance of students trained with traditional pedagogies to new pedagogical methods. For example, in one institution there was resistance to the concepts of "quick and dirty" or rapid prototyping, because students were trained to focus on producing a final hi-fidelity solution directly, without iterations. In another institution junior undergraduate students had significant difficulty with collective problem solving and co-creation in a multidisciplinary environment because they were trained in their secondary education to work individually, competitively, and in each discipline separately.

If the evolution of the organizational culture does not match the academic and organizational changes introduced, there is danger that the achievements are rolled back, sometimes rapidly. One participant called this phenomenon "snap-back", a rapid return to the previous methods and relationships.

Influencing and evolving the organizational culture is a complex long-term undertaking that starts by eliciting the underlying true values and common purpose of the organization. Such an undertaking must therefore be assigned commensurate management attention and the necessary resources with a long-term view for it to achieve the necessary impacts irreversibly.

5.4 Conclusions

The survey data complemented with interviews and discussion depicted a rich and interesting scenario on educational transition towards new paradigms. Overall the emerging scenario was built through a bottom-up process, where leading institutions in the fields of art, design, engineering & applied sciences self-committed to incorporating innovation in their didactical approaches. The set of programs created, while sharing a common vision on breaking traditional disciplinary boundaries, presents highly diverse characteristics. Being in fact a bottom-up process, the observed scenario refers to highly diverse approaches that go from implementing multidisciplinarity into traditional courses to transforming the whole pedagogy of a school. In this perspective transdisciplinarity still remains a prospective goal, only touched by few institutions and not generally informing their entire strategy.

This demonstrates the lack of real and formalized policies at the level both of single institution and of general public bodies. Therefore, these programs are often advertised as points of excellence and are much better known outside their own institutions and are less relevant and influential within their institution. They are usually acknowledged to be the most advanced academic facet of their institution, but at the same time, there is significant resistance to broadly spreading these new models and enabling them to achieve a real impact on academic education.

The lack of resources, constraints from dated national regulations responding to obsolete educational requirements, and internal administrative and cultural resistances made the surveyed initiatives very difficult to implement. Nevertheless, the positive outcomes of these innovative initiatives and their acknowledged success and resilience represent a unique opportunity to reflect on the higher education system and its necessary transformation, to encourage and guide others to initiate their own trans-disciplinary programs, and to possibly inspire new formalized policies.

References

Anderson N (2012) Achieving higher education graduate attributes in the area of creativity, innovation and problem solving through the use of design thinking. In QSApple Conference Proceedings (Vol. 7, pp. 29–33). James Cook University.

Beetham H, Sharpe R (eds) (2013) Rethinking pedagogy for a digital age: designing for 21st century learning. Routledge, New York

Deeun K, Sunhee J (2013) The impact of multidisciplinary design education on the creative process in collaborative design. In: Design learning for tomorrow. Proceedings from the 2nd international conference for design education researchers, ABM Media, Oslo, Norway, 14–17 May 2013

Dimmock C and Goh JW (2011) Transformative pedagogy, leadership and school organisation for the twenty-first-century knowledge-based economy: the case of Singapore. School Leadership & Management, 31(3), 215–234.

Dym C. L, Agogino AM, Eris O, Frey DD and Leifer LJ (2005). Engineering design thinking, teaching, and learning. Journal of Engineering Education, 94(1), 103–120.

Litzinger T, Lattuca LR, Hadgraft R and Newstetter W (2011) Engineering education and the development of expertise. Journal of Engineering Education, 100(1), 123–150.

Ito T, Ichikawa T, Hanumara NC and Slocum AH (2014) Teaching Creativity in Design Through Project-Based Learning in a Collaborative Distributed Educational Setting. In Cloud-Based Design and Manufacturing (CBDM) (pp. 231–282). Springer International Publishing.

Kozma RB (2005) National policies that connect ICT-based education reform to economic and social development. Human Technology: An interdisciplinary journal on humans in ICT environments 1(2):117–156

Kwon D, Jang S (2014) The Impact of Multidisciplinary Design Education on the Creative Process in Collaborative Design: Archives of Design Research 27(2):57–79

Kurokawa T (2013) Design Thinking Education at Universities and Graduate Schools. Science & Technology Trends Quarterly Review 46:50–63

Chapter 6
Innovation Leadership in Action –Today and in the Future

John Body and Stefano Ceri

6.1 Introduction

In this chapter, we look at innovation leadership in action, specifically how the presence or lack of leadership has shaped the innovation process. We concentrate on complex problems and great challenges; indeed, such problems highlight where new leadership is required, as they feature higher dimensionality, and a great number of stakeholders with a variety of objectives and needs, and the decision making cannot be reduced to a top-down technical exercise. We show some situations where the use of innovation as a modality, along the dimensions that were discussed in Chap. 3, is a fundamental ingredient for inducing positive change; and we also show, on the negative side, situations in which the lack of innovation leadership has amplified conflicts and difficulties in delivering results. The discussion on acts and processes can be viewed as a natural complement to the main topic of this book, while the other chapters focus on the education and training of the individuals who will be the major actors of innovation. If these individuals are not integrated in processes leading to innovation (or do not set such processes up), it is unlikely that they will be able to deliver results.

In a turbulent world such as the one we live in, it is becoming apparent that mastering the "innovation game" can be a key determinant of survival and prosperity on all scales, from small organization to large country. In the 2011 State of the Union Address, President Obama made a statement that many business and political leaders would like to underwrite and claim for the entities they are responsible for: "In America, innovation doesn't only change our lives. It is how

J. Body
Founding Partner, ThinkPlace, PO Box 5249, Kingston 2604, Australia

S. Ceri (✉)
DEIB Dept. Politecnico di Milano, Via Ponzio 34/5, 20133 Milano, Italia
e-mail: stefano.ceri@polimi.it

B. Banerjee, S. Ceri (eds.), *Creating Innovation Leaders*,
Understanding Innovation, DOI 10.1007/978-3-319-20520-5_6

we make our living". The problem of stimulating and driving innovation forward is high on the agenda of many corporate and political leaders, despite the frequent frustration and difficulty in obtaining tangible results. Every sector has reason to seek innovation actively, and has to react to an innovation deficit within and outside its organizations. Many organizations are taking a very aggressive role in growing the innovation ecosystem, especially the innovation leadership at the level of both the individual and the organization.

Now, it is quite easy to recognize that innovation is not a "functional problem" that you can solve by simply creating a new corporate function and allocating part of your budget for it. Similarly, you can't just bestow responsibility for innovation to any given organizational entity and hope that this decision will lead to results. Innovation is something that touches on all aspects and units of an organizational or societal entity, most of the time transforming it, and sometimes disrupting it. If we look at it from a scholarly perspective, innovation has too often been reduced to its technological dimensions, and been equated with research and development activity. Today, many scholars still measure innovation at a corporate or national level by counting patents and evaluating research and development investment. This way, they not only fail to grasp the many non-technological aspects of innovation, but also quite dramatically confuse inputs with outputs, generally because it is easier, and it has always been done in this way.

Let us recall the sequence of arguments covered in this book. In Chap. 1 we discussed the types of innovation challenges that we are confronting, in Chap. 2 we looked at the innovation ecosystem where innovation takes place. Chap. 3 introduced the critical role of innovation leaders in shaping, building and mobilizing the innovation ecosystem around complex challenges. Strong leadership is required so that problems are not addressed in isolation but addressed by recognizing the interconnected nature of challenges, with their economic, social and environmental implications. Chaps. 4 and 5 examined how university systems should change in order to train individuals who understand this new notion of leadership, with an overview of programs that are already aware of the required changes.

This brings us to Chap. 6. In this chapter we are taking an applied view. What are examples of complex contexts where new innovation leadership has led to a positive outcome? What problems, where the lack of leadership modalities as discussed in Chap. 3, have created damage? What are some of these problems? Where have we been successful in addressing them? Where have we not? We conclude with a discussion of emerging challenges that will require a heavy application of the innovation leadership described in this book. We start with two cases where innovation leadership has been used positively in action, resulting in a positive evolution of a whole ecosystem. Then, we discuss controversial cases, which exhibit a lack of innovation leadership, leading to economic failures or to social conflicts. Finally, we discuss a case, which is at a very early stage of development, and show how its positive evolution will require innovation leadership at its full capacity; we also discuss the main challenges for the new millennium and call for new innovation management.

6.2 Innovation Leadership in Action

In this section we highlight the positive influence of innovation in producing new solutions for improving developing countries or for solving complex environmental challenges, as well as discuss the forms of leadership that have driven the processes.

6.2.1 Case 1: M-Pesa: An Innovative Payment System for Developing Countries

M-Pesa is a system for electronic money transfer and storage that uses mobile phones; the system was engineered by a remarkable ecosystem that was created after observing how the method was used on a small scale. Figure 6.1 illustrates the main phases of the M-Pesa money transfer model.

a) **Challenge**
People in developing countries often do not have access to simple banking products that people in the developed world take for granted. To move money from one location to another requires the physical movement of that money, for example, walking from one village to another to make a payment. Similarly, the options for storing money can be difficult. When cash is heavily relied on, security and interest are limited (Jack et al. 2011).

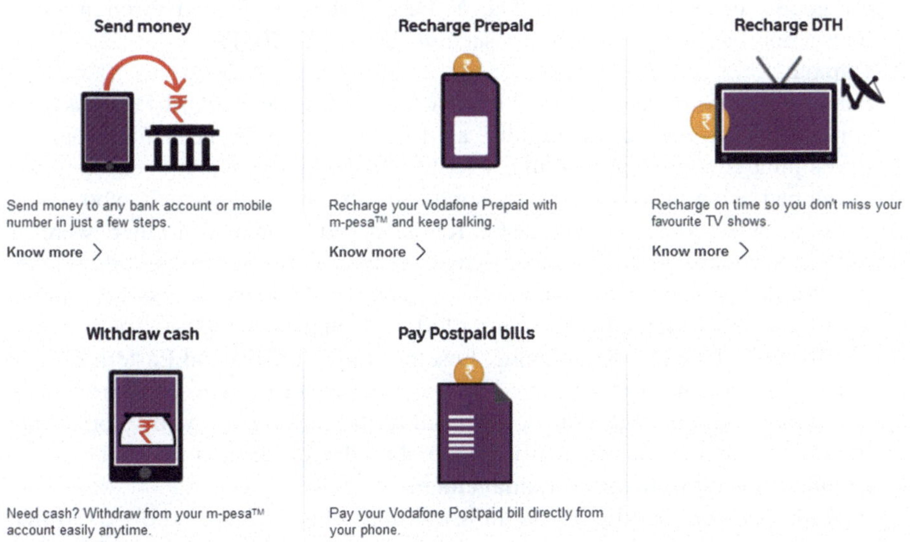

Fig. 6.1 The M-Pesa model for money transfer (from https://www.mpesa.in/portal/services/services.jsp accessed 17 September 2014)

Over the past decade there has been a large number of initiatives to address this challenge using a range of models. One such model emerged after researchers noticed that people were using mobile phone time as a proxy currency. Mobile phone time was stored value that was more portable and secure than money. It could easily be transferred from one user to another. By necessity, people in developing countries had invented a new electronic currency platform (Jack et al. 2011).

b) **Planned outcome**

The planned outcome was to increase the capacity for people in developing countries to be included in the real economy, by giving them an opportunity to trade and transfer funds which increase an individual's overall wealth and capacity, and to have a buffer against unplanned adverse events.

c) **Approach taken**

The innovation ecosystem mobilized around this development. The Commission for Africa connected the researchers who had observed the mobile phone credit exchanges with people at Vodafone, a significant participant in the African mobile telecommunication market. Several industry participants, funding partners, NGOs, researchers and students became connected in an ecosystem around this challenge. The ecosystem provided funds, scale, ideas, technology and program understanding to allow the challenge to be addressed.

The service that was developed is M-Pesa. It is marketed through Safaricom and is a mobile phone-based payment and money transfer system. It was launched in 2007. The service allows people in dispersed geographic areas not served by traditional banking to deposit funds, transfer funds, make payments for goods, or withdraw cash. The M-Pesa system is funded through small transaction fees made on each transaction (Jack et al. 2011).

d) **Impact**

Perhaps because of the ecosystem approach to its development, M-Pesa has been very successful and scaled rapidly, as it serves an unmet need for people in developing countries, and is highly usable. It works in context because it was developed using trans-disciplinary approaches by including people who would use the service. M-Pesa developed in Kenya in 2007, and by 2012 there were 17 million accounts in Kenya alone, where M-Pesa is the largest mobile phone-based financial service available in the region. M-Pesa has successfully spread across countries including Tanzania (2008), Afghanistan (2008), South Africa (2010), India (2011), Mozambique, Lesotho, Egypt (2013) and Eastern Europe (2014). The impact is that people in developing countries who were previously financially excluded can now be included in the real economy, by storing and transferring money electronically via a mobile device (Jack et al. 2011).

e) **Implications on innovation management**

A need was identified by people in leadership roles. It was also observed that people had organically invented a proxy finance exchange system by using mobile airtime as an exchangeable currency. The M-Pesa concept took what was happening naturally and built a financial exchange system around it. Many parts of the innovation ecosystem have been involved. The UN and NGOs have been involved in some of the design work and evaluations. Vodafone has provided the scale for implementation, and the governments involved have addressed regulatory issues. Technology companies have been involved together with financial institutions, and there has been significant involvement from people in developing countries to ensure that the system would work in practice.

6.2.2 *Case 2: Australian Water Rights Trading*

Water management is a complex problem which typically involves many parties, and requires good management and fair regulations. Figure 6.2 shows aspects of the *Water Act 2007* (Austl. 2014), used in Australia.

a) **Challenge**
Australia is the world's most arid continent. Droughts are regular and severe with much of the continent unsuitable for conventional settlement or agriculture. Water is therefore scarce and there are significant debates about getting the right balance of water to sustain local inland communities, meet the needs of agriculture and preserve environmental flows.

b) **Planned outcome**
Australia has relatively few inland river systems. One of the most significant is the Murray Darling system, which runs from the north to the south of the continent on the eastern side of Australia. The planned outcome is to develop an equitable way to share scarce water resources, by managing the overall extraction of water from the river systems, through a transparent process that could be accepted by numerous shareholders.

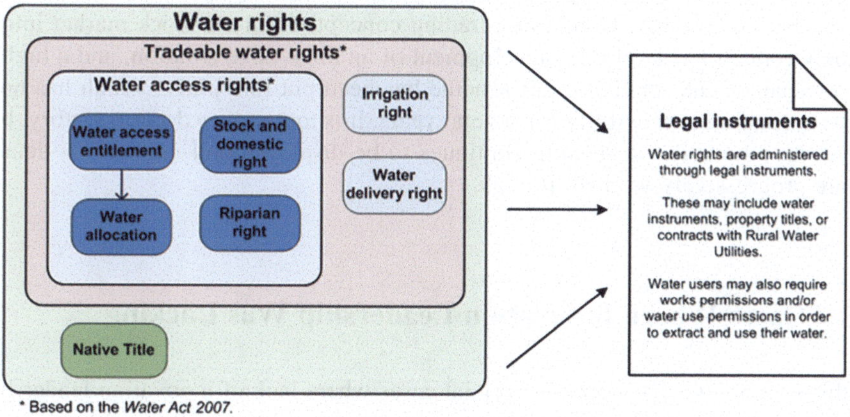

Fig. 6.2 Rights involved in the formation of the water act
From http://www.nationalwatermarket.gov.au/about/rights.html accessed 17 September 2014

c) **Approach**
A government authority, the Murray Darling Basin Authority, has been established to develop and manage the resource. Although this has been a highly debated space, by community, government and industry, progress has indeed been made, with a Murray Darling Basin plan agreed.Working alongside the plan is the *Water Act 2007 (*Cth), which required the basin plan to establish water trading rules that would create an efficient water market. Every liter of water that runs through the Murray Darling system is accounted for with careful metering and measurement. The *Water Act 2007 (*Cth) sets out the need for low transaction costs so that water rights can move inexpensively between parties. Parties interested in the purchase of water rights include communities, government (for environmental flow preservation), and industry (for agriculture, mining and related purposes). The ultimate purpose of the scheme is to have water flow to where it is needed most.

d) **Impact**
The result is the largest water trading platform in the world, yet the solution is not without contention. There are those who argue that water rights trading has increased sustainability and encouraged innovation in the use of water. There are, however, challenges still present and the debate is by no means over. There are concerns, particularly from farmers, that water rights are falling into the hands of multinational firms, which will drive the unit price to a level that may be unaffordable for conventional family farmers.

e) **Implications for innovation leadership**
Environmental challenges are an important arena for innovation leadership. We tend to view natural resources as free to exploit and pollute, but this view is reaching its limits as fast as resources are reaching their limits. When a price is put on a resource that was previously free, it changes the business model and can make what used to be competitive uncompetitive. Despite the challenges, this case shows the leadership that has been demonstrated by government, industry and the community. Combining trading concepts from the stock market into a new context has led to the development of an innovative solution, and a highly transparent and relatively fair scheme has been put into place, which has now been running successfully for several years. It is too soon to declare victory, but strong innovation leadership continues to be displayed and required as details are progressively worked through.

6.3 Cases Where Innovation Leadership Was Lacking

In this section, we discuss controversial cases where lack of innovation leadership may have contributed to increasing the inherent complexity of the problem.

6.3.1 *Case 3: Iridium*

Iridium was established as a company in 1990, in response to the limitations of land-based mobile phone technology. The Iridium case describes the significant failure of an organization, despite having high-profile global investors (most notably Motorola), as well as access to leading technologies (Collins, 2009; Fig. 6.3).

a) **Challenge**
   At the time, mobile phone services often proved unreliable, with service providers lacking the necessary infrastructure to meet the needs of customers. Both land based and satellite phone technologies were in their infancy, and investors were of the view that the global coverage offered by satellite-based phone systems would be superior to existing ground-based phone systems (Lim, Klein, & Thatcher, 2005*).*
b) **Planned outcome**
   The business plan for Iridium was ambitious. The company aimed to use 66 satellites to provide coverage for its mobile network anywhere in the world. The market in particular was intended to be organized beyond large urban centers. This included defense organizations, shipping companies and companies involved in resources. While Motorola was the primary investor in the project, with about one-third of the equity, other companies in related fields as well as various countries invested in Iridium (Finkelstein & Sanford, 2000). Iridium's plan was to construct the phone network, supply handsets to customers, and then operate the network, or sell it to another operator.
c) **Approach**
   After launching the company in 1990, Iridium spent the next seven years constructing its network. Modeling had been done on the viability of the project,

Fig. 6.3 Iridium satellite. From http://en.wikipedia.org/wiki/Iridium_satellite_constellation accessed $32# Oct. 8, 2014

showing that while debt in excess of $2 billion would be required, the project would be viable with approximately 0.7 million subscribers, with the business plan assuming many more than the required 0.7 million.

The project was considered both a technological and commercial endeavor. Commercially it was necessary to attract investment by modeling scenarios; yet technological innovation was considered the main focus of the project (Finkelstein & Sanford, 2000).

d) **Impact**

Iridium filed for bankruptcy in 1999, less than ten years after the company was established. The company had exceeded its projected budget by spending $5 billion on building the technical infrastructure required for the plan (66 satellites and supporting technologies), along with setting up organizational infrastructure to operate and manage the phone network, and deal with customers and accounts (McIntyre, 2009). When Iridium filed for bankruptcy, it had less than 10,000 customers. It needed somewhere between 500,000 and 1,000,000 customers to be viable.

There were a number of reasons why the project failed to attract the requisite number of customers. There were alleged problems with the handsets at the time of launch that detracted from the initial experience of the network. Perhaps more importantly, the land-based mobile phone network had grown rapidly between 1990 and 1999 (McIntyre, 2009).

While Iridium was building its network, land-based companies were experiencing rapid growth. The land-based system offered significantly lower usage costs than the satellite system. Also, given that more and more people were living in densely populated areas where regular mobile services were adequate, the advantages of the Iridium network were not valued. Furthermore, it is alleged that the Iridium phone did not work without a direct line of sight to the satellite, meaning it did not work in cars or buildings (McIntyre, 2009). In short, a premium had to be paid for a service that was unnecessary for most people, especially in urban areas (Finkelstein & Sanford, 2000).

e) **Implications for innovation leadership**

This case study illustrates how a viable project relies on more than successful technology implementation. The Iridium system worked as planned using a web of satellites, yet despite this success, the project failed to be effective. One significant reason Iridium fell short of expectations was its failure to understand the ecosystem surrounding its innovation. The technology worked mostly as planned, but as land-based technologies advanced, they highlighted the shortcomings of this system, which could have been identified through low-cost, user-centered design. Innovation leadership would have highlighted the need for this project to look beyond technological solutions, to realize that innovation requires consideration of the whole system, to ensure that also nontechnical barriers were overcome.

6.3.2 *Case 4: TAV (High-Speed Train) in Val di Susa*

Trains that can travel at a speed above 250 km/h are called "high speed trains" (in Italian TAV – Treno ad Alta Velocità.) Throughout Europe, high speed trains are quite successful in bridging cities (exemplary cases are: Paris to London in two hours, Paris to Brussels in one-and-a-half hours, Milano to Rome in three hours.) The Lyon-Turin TAV project was publicly announced in the 1990s, as a portion of the "Priority Corridor 6" (East–west EU connection, from Lyon to Budapest), more concretely for connecting Paris and Lyon to Turin and Milan. The train line construction has two sections in the Italian and French territory, built by RFI and RFF (responsible for railway management in the two countries), and a critical section across the Alps, assigned to LTF (RFI and RFF control 50 % each).

One of the most significant pieces of engineering in the project is a 57 kilometer tunnel, which crosses the Alps between the Susa Valley in Italy and heads towards Maurienne in France (Greyl et al., 2009; Fig. 6.4).

a) **Challenge**
   There are already existing rail lines crossing the Alps, in particular a line between Lyon and Turin, but it cannot be used because of the curves and altitude gain in approaching the current tunnel under the Alps, which do not allow the gradual gradients required by high speed train services. Thus, a totally new project had to be undertaken, with a high speed line which crosses the Val di Susa region, whose international section between Italy and France requires a new main tunnel of about 50 kilometers for an estimated cost of over 11 billion Euro.
b) **Planned outcome**
   The main aim of the project is to build a line that should be used principally for passenger services during the day (AV) and for freight (AC - Alta Capacità or

Fig. 6.4 Current design of the TAV line connecting Turin and Lyon; the new route shows the 57 km long tunnel and the inclusion of the Orbassano station.
From: http://it.wikipedia.org/wiki/Progetto_di_ferrovia_Torino-Lione accessed on Nov. 3, 2014

High-Capacity) during the night. The other aim is to desaturate conventional lines which can then be used for regional transport and commuter services. The new line will significantly shorten journey times, as its reduced gradients compared to the existing line will allow AV and AC trains to travel between the two countries.

c) **Approach**

The project was seen primarily as an engineering challenge with little regard for the community and environmental impacts of such a large engineering project. In 2006, under a government lead by Berlusconi, the project was managed under special legislation (Legge Obiettivo) allowing for faster decision making, but at the same time bypassing mediation with other stakeholders in the project, including the local government and the residents of the local area (Marincioni et al., 2009). The initial solution required technical adjustments, which led to removing design errors. In the early design, the train line was not going through the station of Orbassano, an interchange node where containers carried by trucks are loaded or unloaded on high-capacity trains. Val di Susa is an area known for picturesque villages, winter skiing and summer alpine excursions, with a strong local identity which became apparent during two previous public constructions, the electro duct and the freeway. The citizens strongly opposed the new project, with No-TAV forming as an Italian movement against the construction of the line. The movement first began in 1995, but it became widely recognized during protests in 2005 and in the following years. Indeed, with a local population of 90,000 citizens, protests attracted up to 50,000 people. Events have included clashes with police and frequent blockades of highway traffic; some of the protesters faced trials and are now in jail, and the movement has been accused of taking radical standing, attracting anarchic disaffected young people. For this reason, the construction site is currently protected by the military, and constantly monitored by the police (Povoledo 2014).

d) **Impact**

The project has been the subject of much criticism because of its cost, the currently decreasing traffic (both by motorway and rail), the environmental risks involved in the construction of the tunnel, and the supposed worthlessness of the new line (airplanes would still, including time to and from the airport and through security, be somewhat faster between Milan and Paris). The No-TAV movement generally questions the worthiness, cost, and safety of the project, with support from studies, experts, and governmental documents from Italy, France, and Switzerland (Greyl et al., 2009). The new line is deemed useless and too expensive, and its realization is criticized for being driven by construction lobbies. The main objections are: (1) the current Frejus rail tunnel has a low level of saturation with no foreseeable increases in the future, (2) the economic feasibility is quite critical, as costs are certain (and very high) while incomes are just projected, (3) there is a danger of environmental disasters, and concerns about health, due to the documented presence of uranium and asbestos in the mountains where the tunnel is supposed to be bored. While the construction is ongoing, there is still uncertainty about the final outcome. On April 9, 2014, the Italian Parliament has voted for confirming the agreements between Italy and France concerning the TAV project, signed in January 2012. In October 2014, allegations about a substantial increase of costs, published by the major economic newspaper in Italy, were later denied by LTF; but the news was sufficient to raise a debate about whether the current government should insist on the project.

e) **Implications for innovation leadership**

This project was conceived very narrowly, with errors in the localization of the train line, possibly due to the urgency of showing progress so as to secure EU funding. Not enough attention was placed on partnership with local governments, which typically involves "local compensation" (e.g. offering new local services) or in communicating the actual risk. The progression of events highlights what can occur when the complexity of the innovation ecosystem is ignored.

The construction of major infrastructure requires much more than political and engineering expertise. This case highlights the need for innovation leadership to broker all the necessary voices in the change, and actively coordinate panels, which should not only make technical decisions, but also involve political and social counterparts, with transparent communication of the outcome of the process. Particularly controversial areas include a strong business case, the environmental perspective, safety perspectives and community engagement.

6.4 Highlighting Future Needs for Innovation Leadership

In this section, we describe a case that is in development; here, we particularly feel the need for innovation leadership, due to impacting new challenges in a wide and complex ecosystem.

6.4.1 Case 5: Next Generation DNA Sequencing

DNA, or deoxyribonucleic acid, is the hereditary material in living organisms. The information in DNA collectively forms the genome, a sequence of individual codes, made up of four chemical bases. Human DNA consists of about three billion bases; more than 99 percent of those bases are the same in all people (Fig. 6.5). The genome is the information for building and maintaining an organism, similarly to the way in which letters of the alphabet appear in a certain order to form words and sentences.

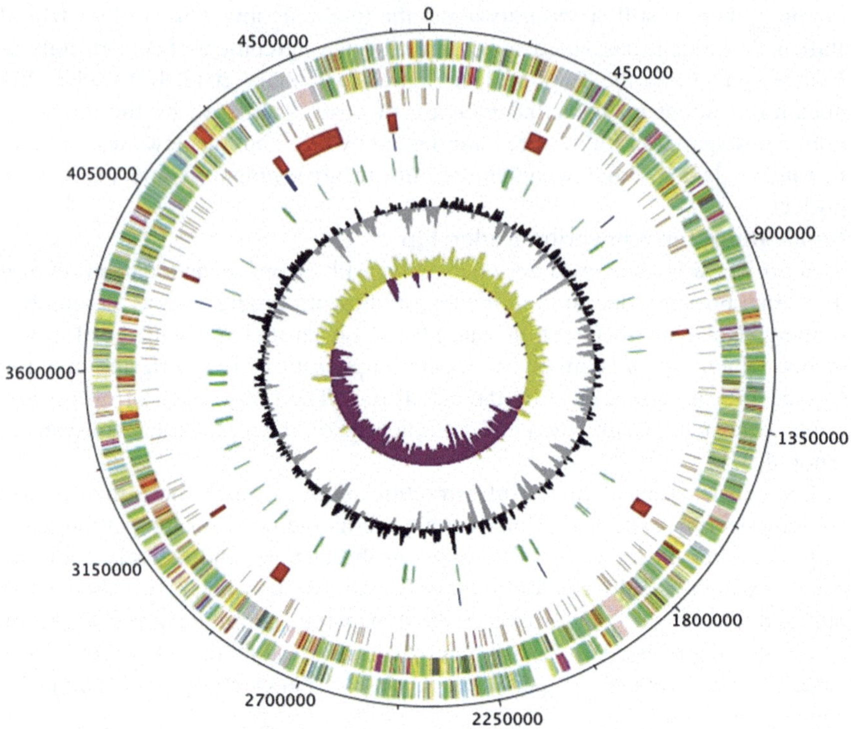

Fig. 6.5 Circular visualization of the genome using a genome plotter
From http://molbiol-tools.ca/Genomics.htm accessed 17 September 2014

a) **Challenge**
A new technology initially developed in 2007, called Next Generation Sequencing (NGS), is increasing the ability to read the genome of all species, including humans. DNA sequencing costs have dropped from millions of dollars at the beginning of the century to about 4,000 dollars in 2014, and are set to further drop down to less than 100 dollars in five to ten years (Wetterstrand, 2014). This will put whole DNA sequencing within the reach of most people, as a new diagnostic and prognostic method. Additional knowledge on biological and clinical properties of the genome needs to be discovered; in many cases, the correlation between genomic information and clinical phenotypes is just probabilistic (Henderson, 2013; Woodcock 2008); yet there is a huge potential for relevant discoveries in the near future, thanks to a greater amount and better quality of available genomic data.

b) **Outcome**
Although the early experiments of NGS technology occurred very recently, a huge amount of NGS data is becoming publicly available (e.g., the ENCODE Consortium data collection, TCGA data about cancer, and the results of the 1,000 Genome project), and large population sequencing programs are either

ongoing or being planned (e.g., in Quebec and Qatar). The pace at which NGS data is being produced indicates exponential growth, which outpaces Moore's law (typically used to indicate the growth of computing power of electronic chips). Data analysis of genomic data is quickly becoming the largest "big data" problem of mankind (Wetterstrand, 2014).

c) **Approach**
While scientists (both biologists and clinicians) are working hard by producing and using NGS technology, policy and legislation is typically slow to develop, much slower than the speed at which science and technology is moving. Unintended consequences may become present long before the law can catch up, and in many cases the law can never catch up (Clayton 2003).

d) **Impact**
There are many consequences that could occur in the near future. Further advances in science may indicate how this newfound knowledge could be exploited for personalized medicine, i.e. for deciding individual treatments. Many new medicines are being developed out of genomic knowledge, including genetic therapies (e.g. for restoring the functionalities of damaged genes due to mutations, or for impacting upon diseases which are regulated by networks of genes). Genetic counseling offers a commercial opportunity to diagnostic companies, which will offer new services relative to specific questions, such as predisposition to diseases or risk of adverse reactions in using certain medicines.

This is already happening, although with controversies (e.g. the case of "23andMe", a company offering genetic counseling, whose activity was suspended in November 2013 by the US Food and Drug administration; on February 2015, negotiations were still ongoing in order to reopen the service, on the basis of the company's reporting proposal). But there are also many unintended consequences, and many more will emerge. If a person can learn at a young age what is likely to happen to him in later life, what impact will this information have? Indeed, the psychological impact of knowing that a certain disease is likely to occur may itself become a problem, with the potential additional difficulty of proper communication, if we consider that most prognostic information is probabilistic and not certain. An ethical dilemma concerns the reporting of incidental findings from whole-genome sequencing, reporting pathogenic variants unrelated to clinical concerns that prompted the testing (Burke et al. 2013). Another concern is relative to the availability of genomic data to insurance companies: will people at high risk of critical disease be eligible for life insurance? Will individuals be eligible for medical treatment if their gene sequencing suggests that treatments may be in vain? Other aspects concern the limits of use of genetic manipulation, and have to do with the ability of changing the genetic profile of people to change their physical or behavioral characteristics.

Finally, there are implications for policy shapers and lawmakers that relate to the information management protocols and safeguards for this data. How much should individual privacy be protected? To what extent can the information be used for law enforcement? What could a government do with the information? Could the information be used for covert intelligence purposes? Could the information be exploited in war or terrorism to target certain ethnic groups? Could information be used to understand public health profiles? Could genomic data be used against individuals, or against the family members of a person whose DNA is sequenced? One has to appreciate that DNA sequenced from one person can reveal much about other individuals in that person's family.

e) **Implications on innovation leadership**
All of the questions listed above were raised in the last decade (Clayton, 2003), but they will need to be answered over the next few years. The quality of the answers will be heavily dependent on the quality of innovation leadership and the health of the surrounding innovation ecosystem. The approach that should be taken requires exceptional collaboration and engagement, and rapid prototyping of concepts with equally rapid evaluation of the prototypes; a number of players are involved, including governments and their administrations, lawmakers, pharmaceutical companies, clinical and biological researchers, and the general population.

6.5 Future Innovation Leadership Challenges

In September 2000, a large gathering of world leaders adopted the UN Millennium Declaration. The adoption committed member nations to take tangible action around eight Millennium Development Goals. The goals set specific targets (Fig. 6.6) and deadlines for achievement. They have been effective in giving attention to specific worldwide problems. The goals are as follows:

1. To eradicate extreme poverty and hunger
2. To achieve universal primary education
3. To promote gender equality and empower women
4. To reduce child mortality
5. To improve maternal health
6. To combat HIV/AIDS, malaria, and other diseases
7. To ensure environmental sustainability
8. To develop a global partnership for development (Fig. 6.6)

Due to the intractable nature of the challenges being addressed by the goals, the actual targets are reductions, but not the elimination, of the listed challenges. Progress has been made in several areas, but the challenges remain as relevant as

Fig. 6.6 Visualization of the eight Millennium Development goals. From http://www.un.org/millenniumgoals/bkgd.shtml accessed on 17 September 2014

ever. Throughout the world's ecosystem, individuals and organizations are looking at how to improve the health of the innovation ecosystem so that the future challenges can be addressed. One example is the growth in innovation labs across the government sector, which is shown in Fig. 6.7.

In a world where rapid human development and rapid advances are being made in all fields to improve lives, the role of innovation leaders is to act. We will need many people with the attributes described in this book to foster the innovation ecosystems to tackle these challenges. Universities should feel obliged to grow these future leaders, with a well-established consideration of the trans-disciplinary implications of such pursuits, and no longer be responsible solely for pursuing deep technical expertise. These future leaders will be deployed into government, industry, the NGO sector and the community more broadly to build the innovation ecosystem capacity for the challenges ahead.

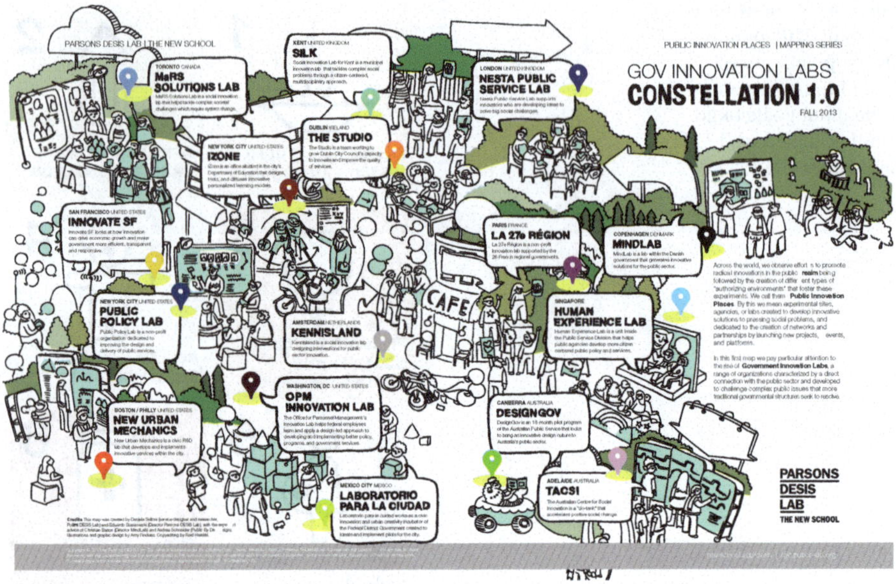

Fig. 6.7 Constellation of Government Innovation Labs.
From http://nyc.pubcollab.org/files/Gov_Innovation_Labs-Constellation_1.0.pdf accessed 17 September 2014

References

Burke W, Antommaria A, Bennet R, Botkin J, Clayton E, Henderson G, Zimmern R (2013). Recommendations for returning genomic incidental findings: We need to talk! Genetics in Medicine, 15(11), 854–589

Clayton EW (2003). Ethical, legal, and social implications of genomic medicine, The New England Journal of Medicine, 349, 562–569.

Collins J (2009). How the mighty fall: And why some companies never give up. Random House Books: London.

Commonwealth Government (2014). National water market. Retrieved September 17, 2014 from http://www.nationalwatermarket.gov.au/about/rights.html

Finkelstein S, Sanford SH (2000) Learning from corporate mistakes: The rise and fall of Iridium. Organizational Dynamics 29(2):138–148

Greyl L, Vegni S, Natalicchio M and Ferretti J (2009). High speed transport infrastructure (TAV) in Italy. CEC Handbook, www.ceecec.net/case-studies/high-speed-transport-in-italy/

Henderson M (2013). Human Genome Sequencing: The real ethical dilemmas. The Guardian.

Jack W, Suri T and Sloan M (2011). Mobile money: The economics of M-Pesa. Retrieved November 5, 2014 from http://www.nber.org/papers/w16721.

Lim J, Klein R and Thatcher J (2005). Good technology, bad management: A case study of the satellite phone industry. Journal of Information Technology Management, XVI(2), 1042-1319.

Marincioni F, Appiotti F (2009) The Lyon-Turin high-speed train: The public debate and perception of environmental risk in Susa Valley, Italy. Environmental Management 43:863–875

McIntyre D (2009). The 10 biggest tech failures of the last decade: Iridium. Retrieved November 5, 2014 from http://content.time.com/time/specials/packages/article/0%2c28804%2c1898610_1898625_1898640%2c00.html

Parsons DESIS Lab (2013). Gov Innovation labs constellation 1.0. Retrieved September 17, 2014 from http://nyc.pubcollab.org/files/Gov_Innovation_Labs-Constellation_1.0.pdf.

Povoledo E (2014) Italy divided over rail line meant to unite. The New York Times, March 17:2014

Progetto di ferrovia Torino-Lion. (2014). Retrieved November 3, 2014 from http://it.wikipedia.org/wiki/Progetto_di_ferrovia_Torino-Lione.

United Nations (2014). Millennium development goals and beyond 2015. Retrieved September 17, 2014 from http://www.un.org/millenniumgoals/bkgd.shtml.

Vodafone (2014). About M-Pesa. Retrieved September 17, 2014 from https://www.mpesa.in/portal/services/services.jsp.

Water Act 2007 (Austl. 2014). Retrieved November 5, 2014 from http://www.comlaw.gov.au/Details/C2014C00715.

Wetterstrand KA (2014). DNA Sequencing costs: Data from the NHGRI genome sequencing program (GSP). Retrieved November 5, 2014 from www.genome.gov/sequencingcosts.

Woodcock J (2008) The human genome and translational research: How much evidence is enough? Health Affairs 27(6):1616–1618

Part II
Best Practices in Higher Education

Chapter 7
Best Practices in Higher Education: An Introduction

Chiara Leonardi and Federica Vacca

The first part of the book has demonstrated the need for innovation leaders and the corresponding impact on renewing education. Specifically, Chap. 5 has discussed a number of experiences in higher education, which are becoming widespread in many universities worldwide. This part is focusing upon the changes that are occurring in higher education, by describing in greater depth some of the most interesting and representative international education programs, which provide concrete answers, although by articulating different approaches, to the need for educational models for creating future innovation leaders.

The common aspects of these programs are the relevance of interdisciplinary experiences and the strong connection established with companies; indeed, companies need new multidisciplinary professionals, able to interact with knowledge and expertise in different areas, from economics to technology and design. On the one hand, this has led the most advanced companies to change the organization of the management processes, making them increasingly more open, parallel and transversal to create and ensure a continuous competitive advantage. On the other hand, it has led the academic world to evolve its educational offering in order to better match the needs of the market, by preparing future leaders for a real process of inter-, multi- and trans-disciplinary innovation.

Each chapter is designed as a structured interview, which entails a defined and ordered set of open questions. Such a design has been followed in most chapters, although with exceptions and deviations, giving rise to structured descriptions that provide information that can be easily compared and can inspire the design of new programs. Although the method appears to be rather rigid, we believe that it has not limited the interviewees in providing us with information about the specificities and

C. Leonardi
DEIB Politecnico di Milano, Piazza L. Da Vinci 32, 20133 Milano, Italy

F. Vacca (✉)
Design Dept. Politecnico di Milano, via Durando 38/a, 20158 Milano, Italy
e-mail: federica.vacca@polimi.it

B. Banerjee, S. Ceri (eds.), *Creating Innovation Leaders*,
Understanding Innovation, DOI 10.1007/978-3-319-20520-5_7

strengths of each institution or program. The structured interview focuses on six aspects:

The Program at a Glance
A form of "identity card" of the program which includes the description of the objectives and activities held, as well as information about the structure and the offered education level.

A Short History
A summary about how the program/course was created, what were the premises, who were the decision makers, what data supported the decision, what were the main obstacles faced in its establishment.

The Educational Mission
A description of the main focus of the program in terms of the topics and keywords with which it is identified and of its educational mission, in terms of specific mindset generation, knowledge, skills acquired, given the background of incoming students. Lastly, it enlists the professional capabilities and profiles developed by the program.

The Community
A summary of the community of faculty and students characterizing the program, with a specific focus on their backgrounds.

Teaching, Learning and Assessment
A description of the educational approach of the program (Problem Setting, Idea Generation, Execution and Implementation, Marketing and Exploitation) and how it is implemented (courses vs. field projects), referencing relevant examples of educational activities and field projects developed.

Feedbacks and Future Evolution
An analysis of the feedback received both by students and external actors such us companies, institutions and recruiters and a description of a possible evolution of the program in the near future.

7.1 The Invited Programs

We invited four North American programs, one program from China, and five programs from Europe; these programs are described in ten chapters that present a wide spectrum of interesting and representative international higher education curricula. North American institutions include Harvard's **SEAS Master Program in Engineering**, Stanford**'s d.school**, Philadelphia's **Kanbar Master of Design in Strategic Foresight & Innovation,** and Toronto's **Master of Design in Strategic Foresight & Innovation**. We also invited the **Mission D** program from Tongji University in Shanghai, and five programs from Europe: Milano's and Torino's **Alta Scuola Politecnica** (**ASP**), Paris-Est's **d.school,** Brighton's **College of Arts &**

Humanities, the EIT ICT Labs **Master School** and the **International Design Business Management (IDBM)** program of Aalto University. All of these programs share a vision of empowering the students through innovation capacity and leadership, which is largely based on multi-disciplinarity, learning-by-doing, and a strong connection with industry.

The Harvard's **SEAS Master Program in Engineering** with its forward-thinking pedagogy is one of 12 degree-offering schools at Harvard University. Established in 2007, SEAS is the newest school in America's oldest university, and is transforming undergraduate engineering education. The School has no departments; most research is interdisciplinary and the curriculum includes significant cross-disciplinary and system-level courses with the aim of reinventing engineering education for the twenty-first century.

The **Stanford d.school** became a hub of innovation and is quickly changing the culture of Stanford. It is not a program but is an Institute that sits inside Stanford, highly engaging the faculty, the students, the industry and the innovation community. The d.school provides to about 750 students coming each year a various and flexible offering, made of classes, pop-up classes, executive education programs and Stanford ChangeLabs experiences, with the aim of experimenting with new and innovative ways of delivering knowledge and experiencing the innovation process. The d.school has created a unique culture of innovation, with studio classes forming the core of the activities.

The **Kanbar College of Design Engineering and Commerce** is one of the three Colleges which constitute Philadelphia University. The curriculum was developed to prepare students to adapt to change, navigate complexity and identify its underlying frameworks, integrate knowledge domains, and identify new opportunities for adding value to the world. Students in the College have individual majors in the fields of design, engineering, and business, but are all required to take a sequential core curriculum that is distributed over four years.

The **Master of Design in Strategic Foresight & Innovation** is a Master program offered by the OCAD University. The mission of the program is to create a new kind of designer: a *strategist* who sees the world from a human perspective and rethinks what is possible; an *innovator* who can imagine, plan and develop a better world. The program interweaves design methods with social science, systemic design, futures thinking and business design with the aim of providing the skills and knowledge to better identify critical issues, frame problems differently and develop innovative strategies, solutions and implementation plans.

Mission D is an interdisciplinary Innovation and Venture Program at Tongji University. Connected with the Aalto Venture Program (AVP), Mission D provides interdisciplinary "design-driven" innovation and entrepreneurship education. Mission D is a Minor Program, which is open to students at the different levels of Bachelor, Master and PhD. The idea is to create an alternative educational opportunity for the students who want to get more knowledge and the capability to integrate and apply knowledge and skills to solve problems in different contexts.

The **Alta Scuola Politecnica** is a school of higher education for the best students of two universities, Politecnico di Milano and Politecnico di Torino. The character of the courses and projects is of multidisciplinary and multicultural nature and focuses on real and relevant problems, proposed as a collaboration between academia and external institutions such as firms, the government or research institutions. ASP projects enable concrete experimental activities on broad and relevant themes and problems. Therefore, selected problems are system-level, interdisciplinary, and focused on problem setting within a complex situation, where innovation plays a substantial role.

The **Paris-Est d.school** is a consortium of five french schools, ENSAVT, ESIEE, UPEM, EIVP and ENPC, with subjects ranging from architecture and urban planning to all types of engineering, along with business and finance. The mission of the Paris-Est d.school "is to become a demonstrator of future pedagogies in innovation thanks to the transdiscipline of design thinking and a worldwide reference in sustainable cities including silver economy, energy efficient buildings and emerging economies." It provides three levels of training, for both master-level students and professors: initiation workshops, intermediary courses, and expert programs.

The **College of Arts and Humanities at Brighton** has been developing and exploring a range of initiatives and innovative approaches to design research and pedagogy under the collective banner of 'Design Futures'. These initiatives are aimed at creating new and responsive educational models informed by the histories and theories of design, and propose a reshaping of the content, mode of delivery, and the means by which these courses will educate designers for the future (GRID educational-developmental tool). As a result, design graduates are able to contribute valuable insights, practices and skills to contemporary society.

The **EIT ICT Labs Master School** is a novel portfolio of Master-level curricula offered by a consortium of leading European universities - a total of over 20 universities, members of EIT ICT Labs, which are driving European leadership in ICT innovation for economic growth and quality of life. The **Master School** reflects an innovative approach aiming to provide flexible, blended learning paths for the new "technical entrepreneur" professional profiles. The mission is to prepare the next generation of highly qualified resources, in both technological and management fields, capable of meeting the needs of existing and future enterprises.

The **International Design Business Management (IDBM) program** is an interdisciplinary offering of the Aalto University based on world-class multidisciplinary and systemic research and learning in global business development through design and technology. IDBM is not a design management program (although is a part of the curriculum), as it has wider scope than that of simply examining the role and practice of managing the design function in a corporate or organizational context. Students originating from different disciplinary backgrounds have distinct worldviews, capabilities and skills that are linked to their institutional backgrounds, and the program has the aim to educate global producers and leaders of innovation in new product, service and business development.

Chapter 8
Embedding Design Thinking in a Multidisciplinary Engineering Curriculum at Harvard University

Fawwaz Habbal

8.1 Summary

The 21st Century inherited challenges that require new methodologies and processes to solve. Engineering has a critical role to play in supporting changes and solving these problems, but our educational system needs to evolve to prepare future leaders to solve and mitigate such human problems. In particular, the educational system needs to shed disciplinary silos and allow students to learn and engage in multidisciplinary dimensions. Engineering embedded in a liberal-arts education provides unique opportunities, especially when design and innovation have been integrated into most of the engineering courses in the curriculum. The case of Harvard School of Engineering and Applied Sciences with its forward-thinking pedagogy is presented here as an example of such a curriculum data and methodologies for assessment of the interdisciplinary courses with embedded design thinking are presented.

8.2 Engineering in a Liberal Arts Context

The School of Engineering and Applied Sciences (SEAS) at Harvard is one of 12 degree-offering schools at Harvard University. It offers to Harvard College students a full undergraduate curriculum, as well as Master's and Ph.D. programs. Established in 2007, SEAS is the newest school in America's oldest university, and is transforming undergraduate engineering education. The School has no

F. Habbal (✉)
Harvard School of Engineering and Applied Sciences, Harvard University, 29 Oxford Street, Cambridge, MA 02138, USA
e-mail: habbalf@seas.harvard.edu

B. Banerjee, S. Ceri (eds.), *Creating Innovation Leaders*,
Understanding Innovation, DOI 10.1007/978-3-319-20520-5_8

departments; most research is interdisciplinary and the curriculum includes significant cross-disciplinary and system-level courses.

SEAS is embedded in a fundamentally liberal arts school. Unlike some programs in engineering and applied sciences, Harvard undergraduates who pursue the field are not admitted to the engineering school but rather are admitted to and remain students of Harvard College throughout their tenure. Concentrators are simultaneously immersed in the liberal arts environment, providing a foundation for understanding the societal context for their technical problem solving.

Moreover, SEAS wants to enable students from all other concentrations to learn how engineering and technology underpin many aspects of society and the world, and thus SEAS courses are open to all Harvard undergraduates. By exposing all Harvard College students to the tenets of engineering – analysis, synthesis/integration, design and building – students gain greater appreciation for science and technology and become better prepared for the 21st century world in which technology is part of every sphere of life.

It is becoming clear that an understanding of changing technology is essential for devising solutions to the world's most wicked problems. The Harvard SEAS curriculum design responds to this need, while also resonating with today's information savvy college students who well understand that the explosion of knowledge and new technology has transformed society as the pace of innovation continues to accelerate.

Against this backdrop, engineering has become essential core knowledge for every broadly educated person – and indispensable background for leaders. At the same time, engineers, scientists, and inventors who will help address the "grand challenges" of the future will need more than technical expertise. In addition to mastering sophisticated new tools and methods from the discipline of engineering, they will also need deep knowledge of societal context as well as critical thinking skills derived from broad exposure to the arts, humanities, and social sciences to affect maximal impact for the common good. Systemic problems like climate change, global demands for energy, cyber-security, clean water deliver, modern infrastructure, and health care for a growing population are not solvable by a single discipline. These challenges are unprecedented in their complexity and require new approaches and methodologies.

Thus, SEAS is reinventing engineering education for the 21st century, and striving to create the "21st century engineer." Our focus is on educating students who excel in engineering and applied sciences, but who also have a broad knowledge of other disciplines, and wish to connect advances in engineering to society's most challenging problems. These "T-shaped" individuals[1] – processing depth in one discipline, but also educated broadly in other disciplines within both the sciences and the arts – are expected to be capable of collaborating seamlessly

[1] This term is used by indicate students who have broad background, yet possess deep knowledge in a discipline. See for example: Joe Tranquillo, "The T-shaped Engineer: Connecting the STEM to the TOP" - 120th ASEE Annual Conference, June 2013.

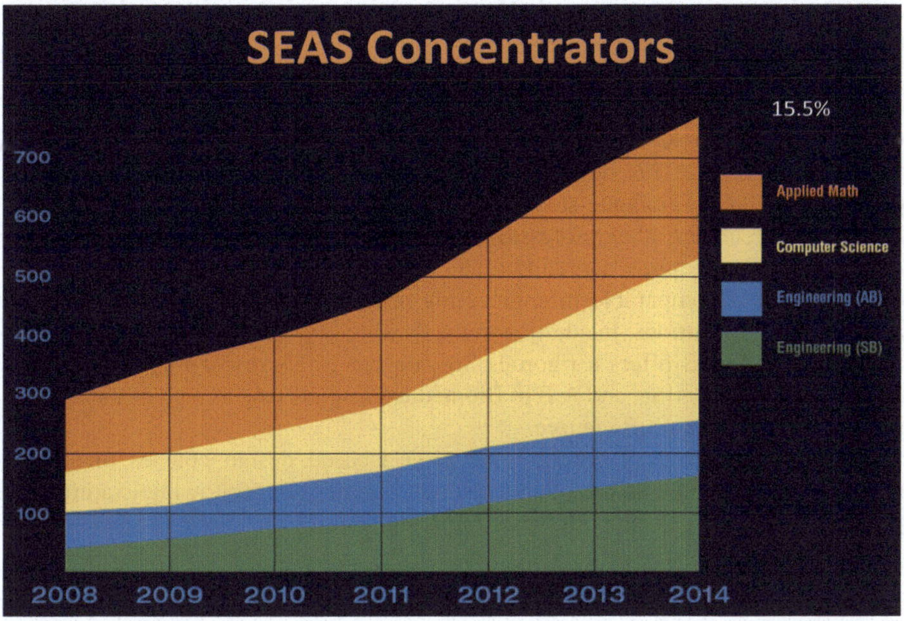

Fig. 8.1 Steady growth in the number of College concentrator enrolled in Engineering and Applied Sciences

across multiple fields spanning arts, humanities, natural sciences and social sciences.

To create this 21st century engineer, SEAS is finding new ways to engage students, deliver content, collaborate across the university, and connect classroom experiences to the wider world. By investing in innovative new instruction techniques and making engineering more accessible to all students, enrollment in engineering courses has increased steadily since the establishment of the school in 2007, and furthermore, concentrators' number has also increased significantly - see Figure 8.1.

8.3 Educational Mission – Active Learning and Design

SEAS is creating an undergraduate curriculum organized around the premise that engineering and the applied sciences are both *multi-* and *inter*-disciplinary. This philosophy leads to a curriculum with a balance of theory and critical thinking skills, as well as deeply integrated hands-on design projects that provide active learning points throughout the curriculum. By emphasizing the *skills of solving problems* by applying iterative feedback to a creative idea, SEAS gives every student an understanding of the design process and the tools needed to solve some of the world's most complex problems.

Harvard is among only a few programs in the US to offer both a Bachelor of Arts (A.B.) degree and an ABET[2]-Accredited Bachelor of Science (S.B.) degree in Engineering Sciences. The *A.B. degree* requires a minimum of 14 to 16 courses for its completion. This degree provides solid preparation for the practice of engineering and for graduate study in engineering, and also is an excellent preparation for careers in other professions (business, law, medicine, etc.). The *S.B. degree* program requires a minimum of 20 courses, and the level of technical concentration is comparable to engineering programs at other major universities and technical institutions. In addition to the flexible Engineering Sciences A.B. and S.B. degrees, SEAS offers a rigorous S.B. degree in Electrical Engineering, and Mechanical Engineering, and A.B in Biomedical Engineering as well as the flexible Engineering Sciences S.B. degree.

The curriculum has a multitude of project-based design courses that teach engineering principles in a multi-disciplinary context. The following examples of project-based courses bring out the different disciplines such courses span:

Computer Sciences 50: Introduction to Computer Science
This course is an introduction to the intellectual enterprises of computer science and the art of programming. Weekly problem sets are inspired by real-world domains of biology, cryptography, finance, forensics, and gaming, and the course culminates in a final project. CS50 is for concentrators and non-concentrators alike, and has the second highest enrollment among all Harvard undergraduate courses (last year 700 students enrolled in this course).

Engineering Sciences 20: How to Create Things and Have Them Matter
Students work in teams to generate, develop and realize breakthrough ideas centered on a theme. The theme varies every year, and past themes have included "the future of water" and "virtual worlds."

Engineering Sciences 21: The Innovator's Practice: Finding, Building and Leading Good Ideas with Others
Students apply a human-centered design process to stimulate innovation, and focus on the interpersonal elements critical for creating and implementing innovative projects in cooperative teams.

Engineering Sciences 22: Design Survivor: Experiential Lessons in Designing for Desirability
Students study real world cases of how organizations strategically design for desirability. This knowledge is then practiced in weekly design challenges, and applied to diverse industries and target markets.

---

[2] The S.B. program in Engineering Sciences is recognized by the national accreditation agency for engineering programs in the United States: 'Engineering Accreditation Commission of the Accreditation Board for Engineering and Technology, Inc. (ABET).'

Engineering Sciences 51: Computer-Aided Machine Design
An introductory course in the design and construction of mechanical and electro-mechanical devices. The course emphasizes hands-on laboratory work using professional modeling software, and culminates in a team-based design project.

Engineering Sciences 52: The Joy of Electronics
An introduction to designing circuits in the context of solving real problems. The course blends instruction with hands-on lab work, and ends with an open-ended project that challenges students to build on core concepts.

Engineering Sciences 139/239: Innovation in Science and Engineering
This course explores factors and conditions contributing to innovation in science and engineering; how important problems are found, defined, and solved; roles of teamwork and creativity; and applications of these methods to other endeavors. Students receive practical and professional training in techniques to define and solve problems, as well in brainstorming and other individual and team approaches. This course is taught through a combination of lectures, discussions, and exercises led by innovators in science, engineering, arts, and business.

Engineering Sciences 159: Introduction to Robotics
This is an introductory course on computer-controlled robotic manipulators. Hands-on laboratory exercises provide experience with industrial robot programming and robot simulation and control.

Engineering Sciences 227: Medical Device Design
A project-based course on the design of medical devices to address needs identified by hospital-based clinicians. Students work in teams with physicians to identify needs and develop a novel device.

In addition to typical engineering courses, SEAS offers *cross-disciplinary design* focused courses. These will be discussed in details later.

The emphasis on design thinking, experiential learning, as well as peer-to-peer learning[3] has permeated across most courses. These elements are integrated within the curriculum and supported by teaching staff and appropriate infrastructure.

For example, a multitude of state-of-the-art rapid prototyping and testing resources are placed in the SEAS Teaching Labs[4]. These labs are staffed by professionals with higher degrees in electrical engineering, environmental engineering, bioengineering, chemical engineering and mechanical engineering. The mission of the Teaching Labs is to provide students with infrastructure and learning using hands-on experiences and tools for problems solving across multiple disciplines. Students are also engaged in skills learning through courses, multiple workshops

---

[3] Eric Mazur, "Peer Instruction: A User's Manual," Addison-Wesley, 2013. Also Derek Bruff, "Teaching with Classroom Response Systems: Creating Active Learning Environments," Jossy-Bass Publishing- Wiley 2009.

[4] See: http://www.seas.harvard.edu/teaching-labs

Fig. 8.2 Students engaged in multi-disciplinary project-based learning

and a multitude of co-curricular and extracurricular design activities -see Figure 8.2.

Faculty members are also engaged in the activities of the Teaching Labs, and most work closely with the teaching staff in designing the appropriate experiments and activities. The teaching staff is responsible for preparing the required infrastructure, whether it is hardware or simulations. Faculty are frequently present during active Teaching Lab periods and work closely with the Teaching Assistants and staff to ensure that students gain maximum educational benefit from the engagements.

The Teaching Labs are also the place where visiting students, from U.S. universities or from other countries, work with SEAS students conducting a variety of projects, some of which are open-ended research projects. These vary in depth and breadth, but all require multidisciplinary problem solving. Examples include dealing with water and air pollutions mitigation, green energy generation, designing medical devices, and developing different types of software projects.

Students are also allowed to use the Teaching Labs for creating devices or executing ideas they have, either as individuals or as part of groups and students clubs. In most of cases, students have mentors from the Teaching Labs. A long list of student organizations at SEAS provide additional opportunities for SEAS concentrators to engage with their liberal arts peers to collaborate on real-world problem solving.

Students have the chance to show their work through an annual *SEAS Design and Project Fair* that is organized by SEAS teaching staff. The fair attracts not only

SEAS concentrators but also those from all across Harvard College[5]. The range of projects displayed every spring is incredibly broad, as dozens of SEAS courses with project components as well as initiatives are represented at the fair.

Many SEAS engineering students choose to increase the depth and breadth of their knowledge by working on extracurricular design projects, either individually or in teams. The goal of these projects is often to implement or disseminate a solution to a problem in the real-world context, outside of the classroom. SEAS encourages students to come up with their ideas and projects that may have commercial value. Students' inventions and related IP are owned by the inventing student(s). SEAS does not share or participate in the ownership of such IP. Furthermore, SEAS offers financial support for these extracurricular projects through *the Nectar process*[6]. Nectar is the official funding process at SEAS to support undergraduate co-curricular initiatives, defined as extracurricular initiatives with curricular (technical) content. Students or groups of students working on co-curricular projects are eligible to apply for a semester funding or longer term funding. Grants for semester projects are typically $2,000 or less, while long-term projects are eligible for a higher funding amount. All students engaging in Nectar projects are required to work with a faculty advisor, and those that require physical prototyping space are often supported by the Teaching Labs. Posters from the Nectar projects are displayed at the end of each funding period.

8.4 Design Thinking and Problem Solving Across the Curriculum

As mentioned above, learning through experiences that incorporate groups have been embedded across the curriculum. Recently, we examined our courses by asking SEAS Faculty to score their courses on percent of design content. Almost 50% of the faculty responded and the result of the survey is shown in Figure 8.3. The survey showed that almost 50% of the courses surveyed have significant (over 50%) design content, and that all disciplines have added design content to their curriculum

The survey showed that almost 50% of the courses surveyed have significant (over 50%) design content, and that all disciplines have added design contents to their curriculum. To better understand the nature of the design content, the Faculty were asked to identify what percentage of the design in their courses is attributed to:

- Problem Solving
- Implementation and verification
- Project management and teamwork
- Communication

[5] http://www.seas.harvard.edu/news/2010/12/es-51-drives-home-principles-engineering-design
[6] http://www.seas.harvard.edu/nectar

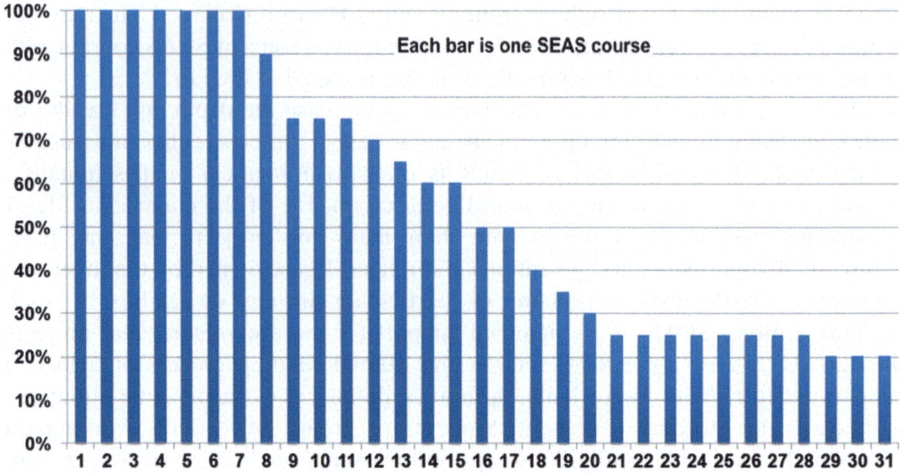

Fig. 8.3 Courses with design content

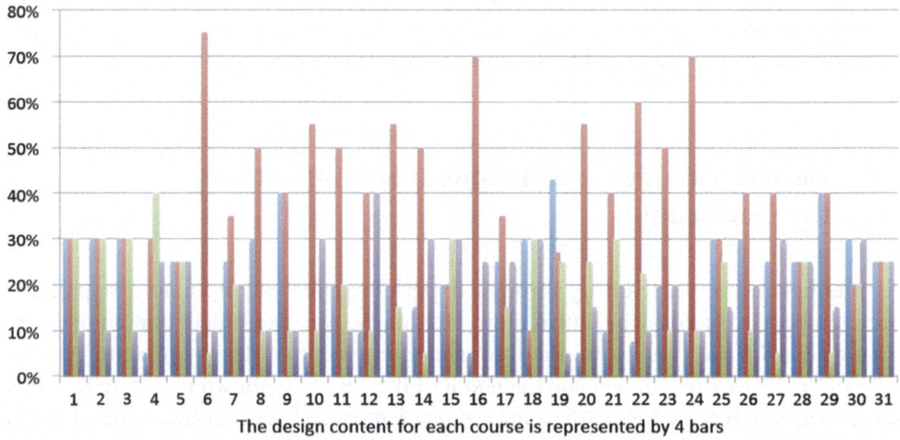

Fig. 8.4 Plotted without a particular order, the design content of each course is represented by 4 bars, providing percent for design emphasis: Problem Solving - blue bars; Implementation and verifications - red bars; Project management and teamwork - green bars; Communication - purple bars

Analysis of this data showed that the design content of the courses participating in the survey has emphasized the implementation and verifications, and then other elements: problem solving, communication and project management. The results of the survey are shown in Figure 8.4.

Analysis of this data showed that the syllabi of the courses participated in the survey has emphasized the implementation and verifications, and then other elements: problem solving, communication and project management.

8.5 Engineering Capstone Courses

In addition to the typical engineering courses, two courses in particular are offered as capstone design courses and have important and complementary goals. These two courses are dedicated to design thinking and problem solving:

1. **Engineering Problem Solving and Design Project**
   (Engineering Sciences 96 – http://es96.seas.harvard.edu);
   Junior year concentrators and non-concentrators ordinarily take ES 96 over a semester. This team project course iterates through the design process to develop a holistic solution to a real world problem.
2. **Engineering Design Projects**
   (Engineering Sciences 100 – http://es100.seas.harvard.edu);
   The design process practiced in ES96 is built upon during the senior year in Engineering Design Project ES100. This course spans a full-year is a requirement for all Bachelor of Science (S.B.) concentrators, and typically executed as individual projects. It provides exposure to a range of technical skills, including performance measurement, quantitative analysis and simulation. Additionally, the course focuses on the user and economic factors that are integral to creating a holistic design solution, such understanding user constraints and needs, problem definition, communication with a client, and documentation and communication skills. It is here where SEAS students' backgrounds in, and exposure to, the liberal arts are critical for creating a cross-disciplinary solution. Students play a large role in shaping this course, from setting deadlines to determining leadership roles to managing group dynamics, thus learning critical project management skills. Faculty and Teaching Fellows guide the students through the design process and provide feedback. The model design process is outlined broadly in Figures 8.5 and 8.6

In a typical ES 96 term session, students work in a group of 10–20 students with a pre-identified client who has posed a particular problem they would like to solve. At the beginning of the course, this problem is only defined as an 'area of opportunity', and it is the students' responsibility to further define and articulate the client's problem and come up with a problem statement. Recent past clients and areas of opportunity have included: improving operational sustainability, with Harvard University Dining Services; using technology to combat gang-related violence, with the Springfield, Massachusetts Police Department; and addressing patient/doctor challenges with managing non-healing wounds in diabetic patients, with a Harvard medical center. Spring 2014 project was on mitigations for the nuclear disaster at Fukushima.

Students must work with their client as one team to understand the overall context, define the problem, brainstorm possible solutions, propose a solution, and prototype and test the proposed solution. Throughout the course they break into small sub-teams as appropriate to the granularity of the problem being solved. Students consult their client regularly to obtain feedback throughout the entire

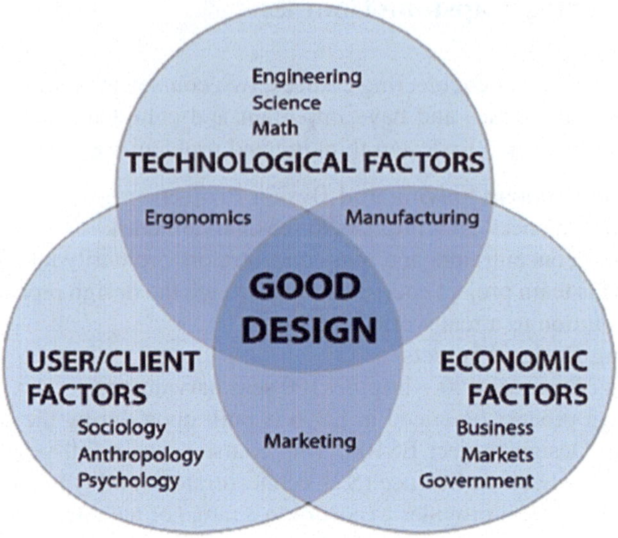

Fig. 8.5 Solutions for real world problems require good designs that are based on interdisciplinary and multifaceted factors

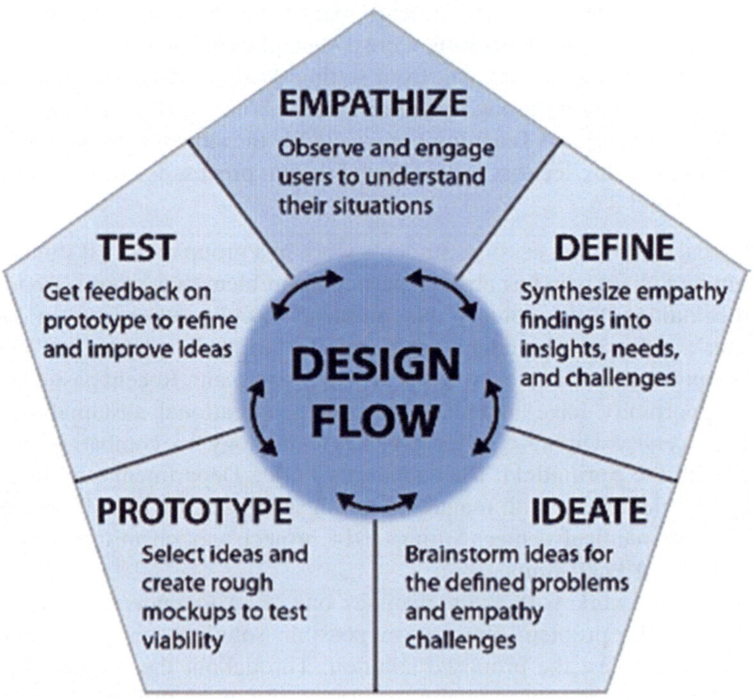

Fig. 8.6 A simplified design flow with interactive team design exercises used to teach students to work within a team and with clients

design process. Over the years, these solutions have ranged from physical prototypes to written recommendations. The course culminates in an hour-long presentation and detailed report presented to the client.

As mentioned, *The Engineering Design Projects* (ES 100) provides a continuation of ES96, but it is an individual engineering design project in which each student will choose and pursue an appropriate capstone project involving both engineering design and quantitative analysis. Each student is supported by a faculty advisor who provides guidance, feedback, and other resources where appropriate.

The range of projects undertaken by the students in a given year is vast. A student concentrating in biomedical engineering may work in a SEAS faculty member's lab to develop a neural-activated ankle orthosis, while a mechanical engineering student may work on building a linkage-based continuously variable transmission. In all cases, the students go through the process of identifying a real-world problem area, refining their problem statement, brainstorming solutions, and prototyping and testing their models. The course involves both the design of the product or system and substantial quantitative analysis to verify and validate that the design meets specified requirements. The course culminates in final individual oral presentations and a final report, which constitutes the student's senior thesis.

8.6 Lessons Learned

One might argue that we are living at an extraordinary time, comparable to the invention of the printing press. Yet, peace, prosperity and equality are elusive as ever, and our world is facing intractable challenges. The complexity of these challenges is very high as they are magnified through complex feedback systems, adding more intensity to their consequences, intended or un-intended. Engineering has a critical role to play in solving human problems, but the structure of the educational system and current pedagogy must be enhanced to enable the emergence of a new cadre of leadership that has the capabilities to engage in transformative interventions. Because these "wicked" problems are very broad in origins and impact, engineering education must become very broad and trans-disciplinary. Since these challenges are rooted in the human system, an immersion in liberal arts education is critical.

At Harvard, the realization that deep technical knowledge is necessary but not sufficient has become part of the institutional philosophy and a cornerstone of undergraduate pedagogy. A focus on understanding complex systems and problem solving is a key to mitigating wicked problems. Education with the purpose of enhancing not only technical skills, but also other human dimensions, including design thinking and mindfulness, is of utmost importance. As a consequence, learning and teaching need to take place in a supportive environment that embraces divergent thinking processes and design methodologies, within an infrastructure that allows for open intellectual exchange, active learning, theory, innovation, and research.

From the discussion presented above, it is clear that one has to deal with a complete ecosystem in order to educate future leaders. Every school must develop and adapt to its own. What took place at Harvard was sparked by the creation of a new school of engineering and facilitated by two consecutive academic Deans, who believed that a paradigm shift in engineering education needed to take place. In addition, the academic environment, characterized by cross-disciplinary, collaborative research, was fertile for making fundamental changes to the curriculum.

The Deans appointed a series of Faculty committees to reassess the goals and structure of the entire engineering curriculum. These committees made recommendations from which interdisciplinary education and design became an integral part of the curriculum. These recommendations led to the hiring of design faculty, establishment of modern Teaching Labs that are supported by high-caliber engineers and technicians, and a new era of support for innovation and entrepreneurship.

Students responded to these curricular improvements: enrollment in engineering programs increased dramatically. On the University level, a multi-schools initiative, initiated by Harvard Business School and supported by the Provost and the President, led to the creation of the i-Lab[7], a new student space to practice and support entrepreneurship. Engineering students were the first to take advantage of the i-Lab.

One may view the past few years as a textbook situation of enlightened leadership knowing how to drive a progressive agenda, and a Faculty believing in the vision and bolstering it. Important enablers of this achievement were a high-caliber Faculty that eschewed academic silos and a trans-disciplinary structure that did not include specific departments. For example, the Faculty's 2008 report on "Design in the Engineering Curriculum" stated that:

> Engineering Design is the central activity of the engineering profession. It is a creative, iterative, and often open-ended process. Its goal is the conception and development of components, systems, or processes to meet practical needs. A designer works under constraints, taking into account technological, economic, and social factors. Engineering design is usually distinguished from other design activities (industrial design, architecture, graphic design) by its use of science and mathematics to provide insight for predicting the performance of prospective designs.

The report goes on to state:

> The great majority of our engineering students follow one of four career tracks upon graduation: enrollment in PhD programs; professional school (especially medical and business school); technical industry; or nontechnical industry. Design is important for all of these and may be the most transferable skill set of an engineering education. The ability to confront open-ended problems and to marshal the resources necessary to address them is fundamental to design education and a key to our graduates' life-long learning skills, both in engineering research and practice as well as in other professions they may pursue.

[7] http://i-lab.harvard.edu

With such conviction, design at SEAS started to be embedded throughout the curriculum and co-curriculum. Yet it was believed that there was additional need to:

> provide exposure to team-based open-ended real-world problems (ES96) and an independent year-long capstone project (ES100) for all SB students....

This turned out to be a critical notion: create cross-disciplinary courses; yet make sure there are courses dedicated to:

- Address messy, real-world, interdisciplinary problems;
- Identify the central needs in under constrained problems and formulate plans to address them (problem setting);
- Explore the extent of the problem space (divergent, creative thinking);
- Develop prototypes as a means of hypothesis testing and exploration;
- Use existing technical knowledge (i.e. engineering science and math) to model and analyze these problems;
- Work effectively on group and individual projects (i.e. personal and interpersonal skills: communication, organization, management, ethics, etc.).'

Another important element that contributed to the success of this program is the presence of mentors[8]. Students are able to seek advice and guidance from professors and from assistant directors of undergraduate studies (ADUS), individuals dedicated to teach and advise. The ADUSs are PhDs in ME, EE, Environmental, Biomedical and Applied Math, and serve as role model and mentors, as well as lecturers of their own courses that emphasize active learning and entrepreneurship. In addition, the Director of Student Life and the Teaching Lab staff serve similar roles. In discussions at focus group meetings, students stressed the importance of their connection to this group of professionals and fondly described the help they receive from them on academic and social matters.

One aspect of preparing engineers to be effective contributors in solving problems is communications and organizational skills. Figure 8.4 above showed the result of the courses, and it pointed out the lack of emphasis on communications and organizational skills. This is not very surprising, as most of Faculty members are concerned with technical content. The dedicated design courses ES 96 and ES 100 emphasize communication and leadership. Such a situation might occur in several institutions, and thus emphasis must be placed on dedicated, high-caliber design courses that incorporate opportunities to develop these important skills.

During the first few years that ES 96 was offered, students were asked about their experiences and, in general, they found the course vague and difficult to follow. They complained about "lack of structure" and "guidance." Drilling down to understand what these comments meant, most respondents indicated that they did not know how to obtain the highest grade and that there were no clear assignments or problem sets to solve. A few years later, most students said that the course was

[8] http://www.seas.harvard.edu/audiences/current-students/innovation-design

the most useful for them in their subsequent work, and many indicated that they learned the most they needed to do their jobs from this course.

Graduate students have requested similar courses on design. Although SEAS offers several design workshops, the feeling was that these fall short of a graduate level curriculum. Recently, the Deans of Harvard Graduate School of Design and SEAS have agreed to offer a joint master degree on design and solving global issues. The design of this program is underway.

Chapter 9
Teaching the Innovation Methodology at the Stanford d.school

Banny Banerjee and Theo Gibbs

9.1 The Program at a Glance

The Stanford Hasso Plattner Institute of Design, also known as the "d.school", is an internationally celebrated hub of innovation. It has been responsible for spreading innovation culture and methods across the Stanford University community, Silicon Valley, and beyond. The d.school has created a strong culture of innovation that places hands-on, Design Thinking-based, team-based studio classes at its core. Central to the approach is its distinct perspective on design: the d.school believes that creativity can be cultivated, not just an innate quality. Anyone can be an innovative designer if they can unlock their creativity and utilize the right process. Its mission is to build creative confidence in every person who walks through its doors, and to make the design thinking method as accessible as possible.

The d.school is an institute and not a degree-granting program, which allows it to provide design education to the entire university rather than a small number of students enrolled in a program. It does not directly admit students nor grant design degrees. Students enrolled in programs across departments and schools apply to take a specific class from a variety of design and innovation classes that are offered to mainly graduate students. The school functions as a dynamic convergence point for students and faculty from across Stanford's academic departments, as well as for external industry and private sector partners. The fact that it does not directly grant

B. Banerjee (✉)
Stanford ChangeLabs, Cordura Hall, Rm. 104, 210 Panama Street, Stanford, CA 94305-4115, USA
e-mail: banny@stanford.edu

T. Gibbs
Stanford ChangeLabs, Cordura Hall, Rm. 110, 210 Panama Street, Stanford, CA 94305-4115, USA

B. Banerjee, S. Ceri (eds.), *Creating Innovation Leaders*,
Understanding Innovation, DOI 10.1007/978-3-319-20520-5_9

degrees is one of its greatest innovation assets—it has a high degree of autonomy, flexibility, and ability to experiment with its own structure and course offerings.

In addition to quarter-long courses, the d.school offers an ever-evolving menu of short "pop-up" courses, skills workshops, and executive education programs that are designed for Stanford students, industry leaders, and entrepreneurs. The d. school also has a year-long fellowship for mid-career innovators.

The d.school offerings are as follows:

Single, Quarter-Long Classes
Approximately 750 students per year participate in the 30–40 annual courses of the d.school. Most classes are by application only and entrance is highly competitive. Classes are ten weeks, some with two or three mini design projects and one capstone project. Most of the classes are meant for graduate students but the d. school is beginning to respond to an enormous demand from the undergraduate community. Courses are usually developed and taught by a team of two or three professors from different departments. Teaching teams often include guest instructors from private sector companies, local design firms, or visiting faculty. Students work in interdisciplinary teams of typically three to five people, and take a d.school class alongside two to four other Stanford classes in their respective departments, e.g., law, engineering, or medicine. All classes are taught in a problem-based studio format. Students learn and apply the human-centered design (HCD) method and work collaboratively to develop innovative products, services, and platforms. The d.school leverages its prime location and the culture of the heart of Silicon Valley to instill a spirit of exploration and entrepreneurship. Many of the d.school courses engage private sector clients to sponsor projects that the students work on over ten weeks. The benefit of these partnerships are two fold: first, the students gain invaluable experience with a real client, often working on cutting-edge technology design projects that open up a range of career opportunities; second, the course project partners benefit from the fresh perspectives and product advancement ideas of a motivated team of high-performing Stanford students. The course sponsorship fees enable the d.school to offer high-quality programming and resources to students and faculty such as travel costs for fieldwork.

"Pop-up" Courses
"Pop-up" Classes are a recent addition to the d.school menu. They are short courses and workshops that range from two hours to two months, and are proposed to the d. school by teams of faculty and external partners. They serve as intellectual and pedagogical experimentation spaces for faculty and teaching teams, and it's not uncommon to see short courses and workshops that are the products of rare interdisciplinary collaborations, such as between a bioengineering professor and a modern dance professor. They also enable greater exposure to design thinking for the student population, particularly to students who have demanding schedules and cannot commit to a full d.school course.

Executive Education
In addition to student courses, the d.school offers intensive workshop courses specifically designed for executives to learn and apply the design thinking process to their companies' challenges. This three-day "bootcamp" is a dynamic mix of short instructional lectures, exercises, and guided, hands-on teamwork. It features a high degree of student-teacher interaction and iterative prototyping and feedback. They go through the whole design process of interviewing and observing customers in the field to understand what motivates them, then generate ideas and prototypes to explore solutions. The in-person workshop sessions are complemented with an action plan for each participant's real work projects and post-program coaching to help participants successfully execute innovation in their own organizations. Aligned with the overall ethos of the d.school, the Executive Education workshops are structured around the belief that anyone can learn and apply design thinking to the challenges they face in their field or company.

Fellowship Program
A recent addition to the d.school, the fellowship is a creative leadership accelerator for early and mid career professionals in everything from education to software design. Fellows enter the year-long program focused on a specific challenge in their domains, and learn how to use human-centered methods to reframe the challenge and tackle it in new ways with the help of an interdisciplinary, cross-sector cohort. A strong emphasis is put on rapid prototyping and leveraging unlikely collaborations to explore new solution approaches. Fellows take courses and also help teach courses and workshops at the d.school.

Innovation Projects
The d.school has several ongoing partnership initiatives to teach and apply design to social challenges. One of these programs in the K-12 lab, where d.school staff partner with primary school educators in the US and other countries to create curricula and techniques to introduce design thinking in their classrooms.

9.2 A Short History

The intellectual roots of the d.school have most recent precedent in the methods of the well-known design firm IDEO, which in turn emerged from the Joint Program in Design at Stanford. Many visionary faculty members have shaped Stanford's unique design philosophy over many decades, including John Arnold, Bernie Roth, Bob McKim, Rolf Faste, and Matt Kahn. Just as the origins of the personal computer and graphical user interface are inextricably tied to 1960s counter-culture around Stanford in the Bay Area, the unusually humanistic turn taken by Stanford Engineering and Design is historically rooted in the same conditions. The Design Program dates back to 1958 when Stanford Professor John Arnold, formerly of the Massachusetts Institute of Technology, first proposed the idea that design engineering should be human-centered. This was a radical concept for engineers in the era of Sputnik and the early Cold War. Building on Arnold's work, Bob McKim (Engineering) and Matt Kahn (Art) created the Product Design major and the graduate-level Joint Program in Design. This curriculum was formalized in the mid-1960's and was one of the first inter-departmental programs at Stanford.

David Kelley, graduate of the Joint Program in Design and later co-founder of IDEO, began teaching design at Stanford in 1978. But the design thinking methodology actually did not catch on until many years later, he recounts. After he earned tenure in 1990, he started teaching classes with different professors from departments such as art, computer science, and business. He found that when students and faculty from the different departments came together, it was easier to come up with innovations because they came with a range of backgrounds, and weren't focused on relying on old habits and structures from their home organizations. "Diversity is the number one thing that correlates to better innovation," he said. Around the year 2000, Kelley and others nurtured the idea of a cross-disciplinary educational Institute for Stanford, as a place to accelerate and evolve the human-centered methodology further, and apply it to new realms of challenges. He mentioned this idea to the Hasso Plattner, founder of SAP and a wealthy industrialist from Germany. Hasso Plattner made a very substantial donation to establish the d.school at Stanford and simultaneously one in Potsdam in Germany. In 2004 the d.school was born as a very small experimental operation. The team of founders were determined to upend the traditional curriculum structure, and create an institute with rapid and adaptive learning cycles—a platform for play, experimentation, and genuine collaboration. "Creativity follows context," says member of the founding team George Kembel. "If I want an organization to behave in a certain way, I need to design for that."[1] As Design Thinking and the human-centered design methodology gained visibility and popularity in industry, the d.school's brand grew quickly. In 2009, it moved into its current home, a building that was completely re-designed to align with the innovation ethos of the d.school. It is distinct from all other learning spaces on the Stanford campus. With movable walls, bright colors, open working spaces, and a seemingly infinite number of whiteboards, the building invites experimentation and fluidity. It emphasises the value of matching the type of task to the spatial affordances, and making sure that the team is optimizing the conditions for the type of thinking that is needed at that point in time. It is a celebration of impermanence and serendipitous encounters and collaboration. "The space isn't precious," says David Kelley. "The whole culture of the place says 'we're looking for better ideas,' not 'keep your feet off the furniture (Fig. 9.1).'"

[1] Linda Tischler, The idea lab: a look at Stanford's d.school, Fast Company.

Fig. 9.1 A Design Thinking Bootcamp in process in the d.school Atrium space. Photo credit: d. school

9.3 The Educational Mission

The d.school focuses on building creative confidence in every student and collaborator who walks through its door. All the courses and programs of the school are structured to achieve this goal. The d.school uses the following key principles to guide its educational mission and programmatic structure:

1. **"Radical collaboration",** across disciplinary and institutional boundaries. Multi-disciplinary teams are a core feature of every class. Fifty percent of the students taking courses in the d.school are from engineering programs; the other half is from non-engineering programs (Natural Sciences, Humanities and Arts, Economics, Business, Law, Medicine, Social Sciences, etc.) It is not uncommon to find a team with a journalist, medical student, lawyer, and civil engineer. Irrespective of their background, students are asked (and trusted) to leverage their creativity and to apply design thinking to real-world challenges facing people around the world (Fig. 9.2).

   "We believe that creative confidence comes from repeated practice using a human-centered creative process to solve problem scenarios called design challenges. We focus on creating transformative learning experiences in which students learn this process together, and then personalize it, internalize it, and

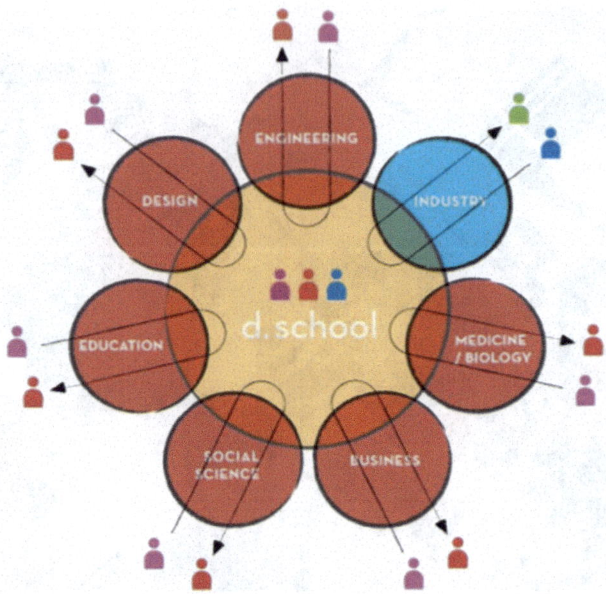

Fig. 9.2 The Stanford Design School's model of multi-disciplinary collaboration. Photo credit: d.school

apply it to their own challenges."[2] A core value at the d.school is to let student interest drive the kinds of projects we work on.

Instead of working on different pieces of the same project, students go through each step in the innovation process together, leveraging their differences as a creative engine.[3] The d.school's physical space facilitates this through providing an open, dynamic workspace where students work together outside of the class (Fig. 9.3).

The d.school culture thrives on the use of rapid prototyping as a methodology, and a bias for action. Contrasting points of view encourage students to see the open-ended nature of innovation and to trust themselves to find their own way forward. The design thinking process creates a container that allows students to take intuitive leaps, explore new ways of looking at old problems, and sometimes start their own entrepreneurial enterprises.

External partnerships (outside of academia) are integral to the principle of "radical collaboration". The d.school partners with corporate, non-profit and government-sector organizations to develop these projects in a learning loop: students get a better understanding of what it means to use design thinking outside the classroom, and the partners deepen their own innovation methodology.[4] During any given academic quarter, there are 50+ sponsored or collaborative projects underway at the d.school. Some are quick introductions that last

[2] www.dschool.stanford.edu

[3] www.dschool.stanford.edu

[4] www.dschool.stanford.edu

Fig. 9.3 Student teams work together in the open, collaborative spaces in the d.school. Photo credit: Silvers et al 2013. http://mw2013.museumsandtheweb.com/paper/design-thinking/

just an hour or two, others are ten-week class projects, and some span years as student teams stick with a project after their class is over. Past and ongoing real-world projects partners include: Facebook, Procter & Gamble, SFMOMA, International Development Enterprises, Kaiser Permanente, Google, Henry Ford Learning Institute, Timbuk2, WalMart, JetBlue Airlines, Mozilla Foundation, and Electronic Arts (Fig. 9.4).

2. **Rapid Prototyping:** Prototyping is central to the innovation and design method that students are taught. It is also a guiding principle to the d.school's overall operations. The d.school spaces and courses are constantly being changed, tweaked, and experimented with. Rather than being a way to validate a solution, prototyping is seen as a way to learn and build better solutions. Students are taught how to use low-resolution materials such as paper, tape, pipe cleaners, and foam to explore and generate ideas at a very low cost. A permanent, well-stocked "costume closet" in the team workspace of the d.school helps students playfully embody the roles of different users or service providers when exploring the dynamics of a product interaction or service experience (Fig. 9.5).
3. **Mindfulness of the Human-Centered Design process:** Intensive focus on process is one of key ways that the d.school is distinct from other design programs at peer institutions. The d.school's educational approach is centered on the belief that diverse team members can leverage their respective expertise if

Fig. 9.4 A range of successful, innovative companies like d.light came out of classes at the Stanford d.school. Photo credit: d.school

Fig. 9.5 A rapid prototype of a human-powered pump being tried out by students. Photo credit: d. school

Fig. 9.6 The image above shows the use of sticky notes in the context of a structured brainstorm, an exercise in rapid idea generation. The term "HMW" refers to "How Might We", a common way of framing a problem as a prompt for ideation. Photo credit: d.school

they are given a common "container" (the design thinking process) with clear steps that everyone can follow together. To this end, the process should be clear and accessible, and every member should be able to understand where in the process the team is. It's critical that everyone be conversant with the process in order to develop innovative ideas effectively. Brainstorming, for example, is a highly generative, "divergent" activity, whereas creating a user's needs statement is a thoughtful, "convergent" one. Course instructors work hard to ensure that teams work in synchronicity in their journey through highly structured activities that help students practice staying in the same process phase together (Fig. 9.6).

4. **Creativity for everyone:** Creativity is something that is learned, practiced, and cultivated. A central, overarching objective of the d.school is to unlock creative potential in people and build what is called "creative confidence." Innovation is partially about the skills and experiences that one has. However, the work of social psychologists and design researchers indicates that the deep mindsets and beliefs that people have about themselves and in their own ability to have creative impact on the world are a definitive factor in creating an innovative thinker. The d.school puts this principle to practice in its affirming, non-hierarchical classroom atmospheres, and through its external programs. For example, it has a K-12 education program to bring the design thinking mindset to the classroom. It also has a leadership program to build creative confidence with mid-level managers at various private and public institutions.

5. **Minimize hierarchy and maximize mutual respect:** There is a high degree of egalitarianism between faculty and student. Classes are "high touch", for both students and teachers. The faculty members spend a lot of time working with students as facilitators and coaches, rather than solely as knowledge transfer agents: meeting the students outside of classes, sharing the same spaces with them, and getting to know each other as people (Fig. 9.7).

In every class, each student and each teacher have a teaching responsibility, a learning responsibility and a doing responsibility. The difference between the students and the faculty is less distinct than in a traditional academic classroom. The d.school pedagogy posits that the students are learning more from each other rather than from teachers' lectures. Students are not seen as empty vessels to fill with knowledge but rather as smart people with complex histories and unique creative capabilities. The job of a faculty member is to create the conditions for the learning process to happen.

As a result of taking courses at the d.school, students' mindsets and skillsets are shifted in several significant ways. Students are more comfortable with the mantra of "failing early, failing often" when approaching a project, and understand the value of low-resolution prototyping. This translates to increased resilience, dynamism, and flexibility in their mindset, which is essential for consistent cultivation of creativity. Students also recognize the importance of interdisciplinary collaboration

Fig. 9.7 A class on "Large Scale Sustainable Transformations" taught by Banny Banerjee and others in progress in Studio 2, one of the highly flexible classroom spaces. The student teams are presenting to each other and are being critiqued by their peers. The non-traditional format helps minimize hierarchical divisions between students and faculty that can stifle authenticity and creativity. Photo credit: d.school

in approaching a complex challenge, and see what that type of collaboration can look like in real-world teams and professions. Overall, students are more confident in their own ability to respond creatively to challenges that are thrown at them—a skill that is applicable to any professional position in their future.

9.4 The Community

There are over 70 faculty members who teach courses at the d.school, and come from every department at Stanford. In addition to faculty, there are external collaborators who often co-lead courses. They come from design firms, private sector companies and even government agencies. Steve Hilton, the chief strategist for David Cameron, for example, recently taught a co-course about Designing Thinking for Public Policy Makers.

The community of students is equally diverse. Students come from all departments and academic levels. The d.school has been able to brand itself as the place at Stanford where curious, self-motivated people go to thrive. With that reputation, students self-select and come to the d.school seeking a new kind of challenge and experience. This self-selection creates a vibrant community of people who are eager to be there and create together—a reinforcing loop between culture and creative output. Everyone who is engaged in the d.school is doing so because they are interested in engaging, and that leads to a high level of passion in the d. school community.

Despite its relative young tenure on campus, the d.school has been able to create a highly visible and growing innovation community. Notably, it has shifted the concept of "innovation" away from its singular association with technology, and rooted the possibility and accessibility of innovation in departments as diverse as History, Psychology, and Natural Sciences. Innovation has become a central principle and goal of the university. Ambitious students feel that their education is *incomplete* without taking a class at the d.school; innovation and design thinking is an expected skill base for a Stanford graduate.

9.5 Teaching, Learning and Assessment

Faculty members have a high degree of autonomy and creative agency. In addition to core, recurring methods classes such as "Needfinding", professors and lecturers from across the university propose classes to teach at the d.school every year. They are encouraged to leverage their own networks in the teaching and learning experience by inviting innovators, intellectuals, and industry leaders from different disciplines and professional worlds to give insight about a particular challenge to students, and serve as guest speakers, coaches and jury members to review students' design products. Classes and curricula are flexible. Rather than deliver consistent, established content year after year—as is the norm in most academic courses—faculty are encouraged to take classes back to the drawing board every year and experiment with something. The goal is to test new frontiers rather than reinforce existing boundaries. Because the d.school does not have any of its own faculty or degree requirements, every instructor who teaches there voluntarily chooses to do so.

Course evaluation and grading are all project-based and are almost entirely defined by each teaching team, rather than by a central administration. In general, assessment is calibrated according to these factors:

- Teamwork
- Degree to which the process was used
- Whether the final product or service was driven by a genuine need
- Creativity and innovativeness
- Degree of completion
- Quality and professionalism of final design work
- Quality of communication and storytelling

9.6 Broader Impact

The d.school has had a deep influence on Stanford, creating a well-recognized platform for design thinking for stakeholders within and outside of Stanford. It enjoys a strong brand, a hub that draws interest internationally from people seeking ways of increasing innovation.

The diversity of classes keep increasing, with many innovators coming to the Stanford d.school to share their insights, and to be a part of the community. Universities and programs the world over are modeling their programs around the d.school structure, and the terms Design Thinking and Human-Centered Design are recognized by a much broader community outside of the design profession.

The demonstration of a crosscutting function in a university that can generate such interest among students and faculty from all across campus has made its culture much more conducive to multi-disciplinary collaboration. The degree of industry collaboration and executive education in non-technical methods have increased. Much like the Product Design program has influenced the engineering program for the past five decades in getting engineers much more sensitive to genuine human needs and to exploring ideas through rapid prototyping.

Chapter 10
Reimagining a University for the 21st Century: The Kanbar College of Design, Engineering, and Commerce

Randy Swearer

10.1 Program at a Glance

Context: Philadelphia University
Philadelphia University has an enrollment of 2,600 undergraduate students and 750 graduate students. The University consists of 39 academic programs recently organized into three colleges with an overall matrix structure: the College of Science Health and the Liberal Arts, the College of Architecture and the Built Environment, and the subject of this section, the Kanbar College of Design Engineering and Commerce (hereafter the College). The University encourages and supports pedagogies that it refers to as Nexus Learning, "active, engaged, real world, and infused with the liberal arts" (Philadelphia University, 2010).

The Kanbar College of Design Engineering and Commerce
The Kanbar College teaches innovation processes to 1,600 undergraduates (primarily between the ages of 18 and 23) in 16 majors. This enrollment represents over half of the university's undergraduates. The curriculum focuses on creating innovation leaders by teaching students to work effectively in team roles, adapt to change, navigate complexity and identify its underlying frameworks, integrate knowledge domains, and identify new opportunities for adding value to the world. One of the premises of the College is that students in design, engineering, and business who do not have these abilities can end up in service roles that offer little opportunity for leadership and advancement.

Students in the College have individual majors in the fields of design, engineering, and business, but are all required to take a sequential core curriculum that is distributed over four years. Core courses commence during the freshman year and build toward an integrative capstone experience during the senior year. The subject

R. Swearer (✉)
Philadelphia University, 4201 Henry Avenue, Philadelphia, PA 19144, USA
e-mail: swearerr@philau.edu

B. Banerjee, S. Ceri (eds.), *Creating Innovation Leaders*,
Understanding Innovation, DOI 10.1007/978-3-319-20520-5_10

matter for core courses is transdisciplinary in focus and explicitly stresses collaboration and teamwork. Course enrollments consist of students from different majors and are designed to foster dialogue across disciplines.

The core sequence is reinforced in the majors with course work, special projects, and industry engagements. The five core courses focus on:

1. Finding opportunities to innovate (Integrative Design Processes)
2. Creating value (Business Models)
3. Managing complexity (Systems Analysis)
4. Observing the world to ask the right questions (Research Methods)
5. Integrating learning in real-world conditions (Capstone Experience)

The Central Role of the Liberal Arts

The Core was designed to be far more than a professional training program. It was always conceived of as a powerful means of teaching students to critically think about the world and their places in it—and in that sense shares common values with the liberal arts. Yet the Core aspires to teach students not only to think critically, but to act critically as well. It encourages what Aristotle called the art of practical reasoning—or finding meaning in the world, and the basis for wisely acting in it, by making education relevant to *particular* experiences, places, and moments.

For all of these reasons, the University's liberal arts program is integrated with the Core; in fact two of the Core courses, Systems Thinking and Research Methods, count as liberal arts course requirements. The liberal arts program was comprehensively redesigned based on four core value propositions as the College was being developed and implemented. These value propositions are strongly synergistic with the Core:

- Question, based on curiosity and confidence
- Adapt, based on contextual understanding and global perspective
- Contribute, based on empathy and collaboration
- Act, based on initiative and ethical reflection

The value proposition statement ends by declaring an underlying purpose resonant with the Core, which is to teach students to, "imagine and realize better futures" (Philadelphia University, 2014).

The College's Executive Dean, Ronald Kander, envisions the Core as a strand of DNA (Kander, June 11, 2012). One spiral forming the double helix represents the academic majors; the other spiral represents the liberal arts. The twisting ladder-like structure connecting each spiral represents the Core sequence (Fig. 10.1).

Fig. 10.1 Ronald Kander's core as strand of DNA

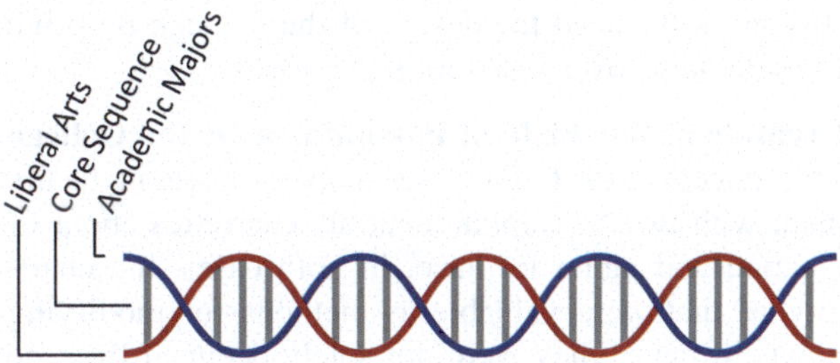

10.2 A Short History of the Program and How It Was Created

A central initiative of the University's strategic plan was to create a College of Design, Engineering, and Commerce (C-DEC) by combining three existing schools. However, the College itself did not originate from the strategic plan's working groups; rather, it appeared next to Initiative 2 on innovation, "achieve innovation by creating a College of Design, Engineering, and Commerce." (Philadelphia University, 2009). The strategic plan's working groups had identified innovation as a central part of the University's culture, but University leadership had reframed the concept of innovation in the form of a new college. The concept of innovation was now something concrete that the University would build together.

Making the College of Design Engineering and Commerce

The College took shape as it crystallized around models and prototypes that the Provost's Office became more adept at using to advance the project.[1] Nearly every initiative for creating the College was positioned as an opportunity to create prototypes for collectively accruing knowledge that would contribute to its design. The first 18 months of the initiative could be read as the process of simultaneously defining the College while implementing it in iterative loops of learning.

The development of the College cycled between a planned artifact that required formal processes and procedures to implement and an evolving prototype that spurred dialog about what it should be. This interplay allowed faculty and administrators involved in building it to continually generate new meanings by confronting their tacit assumptions about what the College might become. There were many deeply embedded assumptions about the College. For example, after the C-DEC was announced, but before it took shape, a few members of the faculty described their interpretation of the C-DEC as an academic unit in which students would "find it [design], build it [engineering], sell it [business]" (McGowan, H., Personal Communication, 2011) Left unexamined, this tacit framework might have

[1] Creating the College was a broad-based university effort, but I wish to recognize the extraordinary work of Vice Provost Gwynne Keathley and Assistant Provost Heather McGowan. Their strategic leadership directly contributed to the success of the project.

strongly influenced the design of the College by privileging conventional knowledge domains over innovation processes.

Creating a New Field of Possibilities for the College

The concept of the C-DEC was made concrete in the first months after its announcement with two "discipline-neutral" charrettes about future forms of news content creation and aging in place. In both cases, the charrette process generated prototypes that were valuable as methods for modifying entrenched cultural viewpoints, because they were not solely deduced from normative assumptions about what a college should be. This point is crucial: the prototyping process encouraged thinking that interwove modes of thinking that were inductive, deductive, and abductive. The innovation process that resulted in the College was most effective when its activities required the community to blend these three ways of thinking. For example, the prototyping process that emerged from the charrettes encouraged "building to think—rather than discussing, analyzing, or hypothesizing in abstract terms before acting" (Coughlan, Suri, & Canales, 2007).

Far more valuable than the design proposals that emerged from the charrette processes was an increased ability by the community to envision new possibilities for the College. The charrettes encouraged the process of transcending the legacy paradigm of "find it, build it, sell it" and adopted a more integrative disciplinary sensibility. In fact, one of the most important results of the charrettes came later in the form of an interactive diagram that demonstrated common C-DEC academic outcomes by year, which led to the concept of interdepartmental course streams (later referred to as pathways) that could unify and buttress the C-DEC academic core courses across disciplines.

Managing the Change Process

The University had begun an action research project (Antheil & Spinelli, 2011) when the strategic plan was first launched to assess the perceptual gaps between the University's current reality and where faculty and staff thought it should be. Action research was a means of assessing change processes as they were occurring by using interviews and survey instruments. The University's action research process focused on issues such as perceptions of short-term versus long-term planning, risk taking versus risk aversion, and autocratic planning versus participatory planning.

Based on this user-centered research, important planning directions were changed; for example, the timing for announcing the organizational structure of the C-DEC was directly related to research indicating that the initiative was in danger of losing momentum, because faculty and staff could not concretely "understand their identity in the new structure" (Antheil & Spinelli, 2011, p. 29). Soon after the announcement, it became clear that the community members actually needed to see their names in a diagram of the College to concretely situate themselves. The resulting organization chart became another powerful prototype, because feedback from the deans and faculty allowed for iterations that directly affected the final organization of the College.

The imperative to maintain momentum resulted in an early announcement of the new building for the C-DEC in the spring of 2009. The building development process was designed as a structured prototyping opportunity that involved the entire campus, including a key faculty planning committee. Schematic diagrams, sketches, and models revealed a number of tacit assumptions about what the College would become; for example, how and where courses would be taught, the role of individual academic units, the role of research, how faculty and students would use spaces outside the classrooms, etc. The architect's prototyping artifacts acted as learning repositories for crucial discussions about the nature and organization of the College.

Building the Curriculum

The curriculum and new building were simultaneously developing with knowledge from one process sometimes feeding into another. In a dedicated room in the basement of a dormitory, faculty, deans, and others used sticky notes to visually represent curricular sequences, which produced a series of four core courses with an integrative capstone. These conceptual kernels were then significantly further developed in multidisciplinary faculty committees chaired by the Vice Provost as the building emerged from planning discussions.

During this period, the vision emerged of an integrated four-course sequence followed by a capstone taken by all 16 C-DEC majors in mixed, team-taught courses starting in the freshman year. The first course was called Design Process and Integrative Thinking (later changed to Integrative Design Process). It focused extensively on team dynamics, opportunity-finding methods, iterative problem-solving, and basic field observation. Sophomore year, students would enroll in Business Models, a course focusing broadly on value creation (economic, social, etc.) Junior-level C-DEC students would enroll in Research Methods, which introduced ethnography and other basic field research techniques. The final course before the capstone was Systems Thinking, with options in biomimicry and sustainability (and later material properties). The course introduced students to systems principles, systems dynamics, and their applications. During the senior year, a major capstone project with external for-profit and nonprofit organizations was designed to integrate the core sequence with the other coursework students had taken in their majors.

Over time the C-DEC core was to be complemented and extended by interdisciplinary course pathways and what became locally known as credential clusters, planned as knowledge ecosystems in areas such as fashion (defined broadly to include a range of consumer products) and healthcare. This process continues today as the curricula in the School of Business Administration and the School of Design and Engineering evolve in ways that reinforce the C-DEC core.

The first C-DEC core course, Integrative Design Process, was offered in temporary facilities in the fall of 2010 as the new building was being constructed. The course building process involved over a dozen faculty members from the College and College Studies (liberal arts and sciences). This group became the nucleus for the first teams that taught the course.

10.3 Educational Mission

Why the College Was Created

The Kanbar College initiative reflected an evolving macro-level perspective at the University about the mismatch between the legacy educational models of higher education and the transformation of professional work in this century. This perspective became embedded in the Strategic Plan as the objective to "become the model for professional education in the 21st Century" (Philadelphia University. Strategic Plan, 2009).

The professional world our graduates were entering had become progressively more dynamic due to macroeconomic trends in recent decades. A prolonged period of deindustrialization, increasingly fluid capital markets, technologically distributed work environments enabled by new communications technologies, and globalization trends were factors that had shaped an economy that was in many senses continually restructuring. These forces, among others, had accelerated the rate at which professions were born, had died, and were evolving. It seemed clear that legacy university models were not designed to prepare students for this new reality. In short, the university saw an opportunity to add significant, differentiated value to professional education in a crowded, commoditized higher education market that was, and still is, geared toward producing graduates for professional trajectories more typical of the industrial era.

Yet, beyond this competitive strategy and the immediate, instrumental, vision of a new kind of professional education was a basic belief that the traditional academic dichotomy between applied and unapplied forms of education had become false and destructive.

Mission Objectives and Outcomes

Kanbar College students learn to create value in the world by applying innovation processes to finding problems that matter and solving them with elegant human-centered solutions. In order to innovate, students gain disciplinary depth in their majors and transdisciplinary breadth in the College core curriculum. This objective is supported by seven learning outcomes.

One of these outcomes is fundamental for driving innovation and is the conceptual bedrock of the Core curriculum: the abilities of students to integrate knowledge in new ways in order to find new opportunities and create new value. Directly or indirectly, all Core outcomes contribute to and result from it. Three additional Core outcomes focus on learning that provides students with a meta-perspective that facilitates their abilities to effectively collaborate on multidisciplinary teams. The Kanbar College curriculum is specifically designed to help undergraduate students learn how to optimize their participation in collaborative environments. This involves learning to describe the value of different problem solving and decision making styles. Effective teamwork also requires students to build a meta-awareness of their own disciplines and those of their teammates in order to identify the unique contributions and limitations of specific disciplinary perspectives.

Another Core outcome focuses on empathy and ways of understanding the human condition from the points of view of those who experience it in specific contexts. This ability requires students to gather insights from people, their behaviors, and their cultural practices. The human condition is endlessly complex. In order to make sense of observed human behavior, students need to navigate complexity in part by having an ability to evaluate the ways in which natural and man-made systems (technical, political, social, cultural, economic, etc.) shape, and are influenced by, new products, processes and services. Students must use this same ability to understand how the solutions they develop will dynamically interact with natural or manmade systems.

Finally, the broad objective of teaching students to understand the world in all of its richness and complexity extends to the professional world into which they will graduate. The Core is about teaching students to understand their discipline and correlated professions as existing on a constantly changing continuum of challenges and value creation. This dynamic work environment will generate rapidly shifting and evolving professional challenges. In order to foster this understanding students must learn to adapt behavior in response to continually changing professional challenges.

Professional Capabilities

Early conceptual blueprints for the College imagined students who would be immediately prepared to take on entry-level professional roles primarily due to the professional knowledge represented by their majors. The transdisciplinary knowledge represented by the Core would provide them with the tools and frameworks to rapidly advance to cross-sector team and project management roles within the first few years of graduation. Initial leadership roles such as these would become the platforms from which they would advance into innovation-driven leadership roles.

The College will produce its first class of four year graduates in the spring of 2015 and therefore does not yet have data on postgraduation professional outcomes. However, there is abundant anecdotal qualitative evidence that the DEC learning experience has had a significant impact on students. Many members of the faculty report that upper class students in the College are much more adept at collaborating and exhibit little of the disciplinary chauvinism that can be so caustic for teams. Teams appear to more quickly uncover the problems that matter, more fluidly iterate solutions by synthesizing discipline inquiry methods, and frame solutions in ways that often make us look differently at the subjects they address.

The growing capabilities of College students have been a key factor behind the significant increase in industry-sponsored projects. This success supported the development of a greatly expanded and integrated industry engagement platform, the Nexus Innovation Program. These projects have consistently produced impressive results, which has resulted in a number of repeat industry partners. Student involvement in the Nexus Innovation Program has frequently led to internships and job offers for graduates, and directly contribute to the University's 94-96 % employment or graduate school acceptance rate. The tremendous success of the

C-DEC-driven Nexus Innovation Program earned the University a major award from the University Economic Development Association (UEDA) for Excellence in Innovation and Entrepreneurship.

10.4 The Community

On the simplest level the community consists of students, faculty, staff and administration in the College of Design Engineering and Commerce. Yet the success of the College is based on a much larger ecosystem of communities consisting of other colleges (and faculties) at the university, networks of external partners, DEC Fellows (official outside advisors with special expertise), and philanthropists. The effort was also developed and executed with a substantial infrastructure of leadership, consultation, and support from the Provost's Office and its constituent units. The C-DEC student experience is dependent on aligning these communities with its mission, and orchestrating them at the right time and with the right intensity to deliver the academic program.

10.5 Methodology

The Core should be imagined as a spiral of knowledge that progressively introduces and reinforces methodologies at higher levels of sophistication.

The problems students confront in the Core are rarely simplified for pedagogical purposes, even at the freshman level, because the program puts a premium on exposing students to volatility, uncertainty, complexity and ambiguity (VUCA). The curriculum itself is designed to help students navigate VUCA problems and find opportunities in them for innovation.

Problems are based whenever possible on real world conditions. Depending on the course or sponsored project, they might concern the supply chain of a major corporation, use of a library, delivering social services, developing new approaches to wound care, or creating a new generation of urban rooftop wind generators. Student responses to these problems, in the forms of analyses, models, prototypes, and solutions, are more or less sophisticated depending on how advanced they are in the curriculum.

VUCA problems create the need for collaboration. In order to understand and exploit them for innovation, students need to tap into the "collective brain" of their interdisciplinary teams. The program provides extensive knowledge to students about teamwork and collaboration in areas such as team dynamics, learning styles, and personality inventories, but VUCA problems create the necessity to apply that knowledge and synthesize it for use in disparate contexts.

Integrative Design Process
Students learn early in their Core experience that finding innovation opportunities in real-world VUCA environments requires special research methods. During the first course, Integrative Design Processes, students are exposed to simple field observation methods that include generating empathy with those they are observing, distinguishing between inferences about what is happening in the environment and inductive analysis of the field data, and avoiding sensory bias. These methods are repeatedly exercised during the College experience by courses in the majors, interdisciplinary projects, which often include academic areas from outside of DEC, and a wide range of industry (for- and non-profit) projects.

Business Models
VUCA problems require students not only to navigate complex data, but also to use it for generating new kinds of value. Therefore, the student toolkit of research methods expands in the second Core course, Business Models, which focuses on value creation. The research methods in this course are oriented toward understanding the components and processes of value creation across financial, cultural and social dimensions. The course uses a business model framework influenced by the canvas approach presented in the book by Osterwalder et al. (2010).

Systems Analysis
Research methods continue to spiral into a third Core course, Systems Analysis. This course is highly synergistic with Business Models since the frameworks students learned in that course are in fact systems for value creation. Students in Systems Analysis learn to navigate VUCA environments by learning about stocks, flows, and other key systems components. They use a systems framework to model actual organizations, natural and man-made systems. There are three courses offered for Systems Analysis, all of which count as a general education requirement: Biomimicry, Sustainability and Materials Analysis.

Research Methods
Approaches taught in the first three Core courses for understanding VUCA environments are deepened and broadened in the forth course, Research Methods. The course, which also satisfies a general education requirement, uses a social science lens to add depth and discipline to a student's understanding of field research methods, including ethnography. The knowledge spiraling through this course is particularly powerful for building on data collection methods learned in Integrative Design Process.

Capstone
The Core capstone occurs during the senior year. It is specifically designed to demonstrate the abilities of students to understand and apply the Core learning outcomes to major projects. These projects are often from businesses and organizations outside the University and collaborative in nature.

Assessing Learning in the Core
A team of faculty and the Provost's Office developed a survey instrument to assess the four year Core learning experience and its constituent courses. The survey was specifically designed to measure the degree to which students were learning the formal Core outcomes. As part of the survey development process, the team identified a number of specific student behaviors that would support the outcomes. For example, these behaviors included, "apply material learned in one course to other courses in your program," "work on a course assignment that requires integrating ideas from different disciplines and sources," and "recognize that changes in one area impact other areas of inquiry" (Philadelphia University. College of Design Engineering and Commerce, 2013). A key group of these behaviors emerged from the first annual review of the curriculum that strongly correlated with successfully meeting learning standards for all seven Core outcomes. Ongoing iterations of the Core and supporting activities such as industry sponsored projects focus on developing this set of fundamental learning behaviors.

10.6 Lessons Learned

The College Core experience was designed for undergraduates primarily between the ages of 18 and 23. This focus resulted from a variety of strategic factors at the University, but was driven by the belief that by providing an innovation curriculum at the undergraduate level students would be less likely to develop unproductive disciplinary biases, be better able to put their majors in a broader framework of disciplines and professions, and be socialized and acculturated in ways that promoted inter- and trans-disciplinary teamwork and collaboration.

Many of the implementation challenges of the College Core resulted from issues pertaining to the strategic focus on this age group. For example, freshman have typically attended high schools with highly structured curricula stressing individual rather than group performance, and structured problem solving oriented toward right or wrong answers, and have very limited understanding of disciplinary and professional pathways. Annual iterations of the curriculum have addressed many of these issues, especially at the freshman level.

10.7 The Future Evolution of the Program

The principles and cultures of innovation represented by the Core have had direct and indirect impact on the university as a whole. Innovation represented by the College and its curriculum has increasingly become central to the University's identity, driving decisions about coursework, facilities, and new programs. Curricula in the College and across the campus have been created or modified to reinforce the Core—or extend its reach to new subjects and methodologies. On a broader cultural level, several members of the College leadership team point to an increased general willingness by faculty to experiment with curricula and new program structures. From the vantage point of the Provost's Office the openness to deepening the culture of innovation has extended well beyond the College to many areas of the University. A concrete example of this influence is the fact that the College of Architecture and the Built Environment has integrated aspects of the Core into its undergraduate program.

The Core has also helped to attract outside funding such as the Blackstone LaunchPad, which fosters entrepreneurship by working with students to develop concepts into viable products and services.

The Multi-Modal Evolution of the Core

Although the Core was originally targeted at undergraduates, it was (like many of the University's new academic programs) always designed to flexibly shift shape and subject matter for a variety of contexts and audiences. This flexibility has furthered the University's vision of its future students as being distributed more evenly on a continuum consisting of populations that are pre-college, undergraduate, graduate, post-graduate, and professional (business, not for profit, government). The Core has been translated for a variety of audiences; for example, the University partnered with the City of Philadelphia to create a Municipal Innovation Academy that delivers Core principles and pedagogies for promising city managers positioned to drive change in government. The University has also adapted the Core to spur innovation by employees in a variety of industries such as healthcare. Pre-college audiences have also been exposed to the Core in intensive workshop and recruitment events.

The Strategic Design MBA

An excellent example of program innovation at the University that shares DNA with the undergraduate Core, but serves an audience that could not be more different, is the recently launched Strategic Design MBA. This degree program focuses on many of the issues introduced in the Core, but on a more advanced level and deeply infused with business knowledge. As with the undergraduate Core, students in the Strategic Design MBA program also study across a variety of advanced courses, integrative design processes, systems analysis, research methods, and business models. The audience for this program typically has full-time professional commitments, is between 25 and 55 in age and often has families. These students come from backgrounds in non-profit, for-profit, and government sectors. Many are successful entrepreneurs. Most of these students are seeking a transformational and catalytic experience that incorporates, but transcends, the typical, linear and quantitative orientation of normative MBA programs.

The future of the Core and its influence at the University are dynamic. The Provost's Office expects that it will be continually modified and iterated based on College and University assessments and broader environmental conditions. Our goal is to use this process of continual change and renewal as an opportunity to integrate innovation processes, where appropriate, into curricula across the University.

References

Antheil, J. H., & Spinelli, S. (2011). Action research to support the sustainability of strategic planning. Planning for Higher Education, 39(4), 23–31. Retrieved from ProQuest Research Library database.

Coughlan P, Suri JF, Canales K (2007) Prototypes as (design) tools for behavioral and organizational change. Journal of Applied Behavioral Science 43(1):122–134. doi:10.1177/0021886306297722

Kander, R. G. (June 11, 2012). An adventure in extreme curriculum integration to stimulate innovation and collaboration. 119th ASEE annual conference & exposition, session M454, paper: AC 2012–5418.

Osterwalder, A., Pigneur, Y., & Clark, T. (2010). Business model generation: A handbook for visionaries, game changers, and challengers. Hoboken, NJ: Wiley. Retrieved from /z-wcorg/

Philadelphia University. (2009). Vision and strategic plan. Retrieved January 4, 2012, from http://125.philau.edu/vision.php

Philadelphia University. (2010). Philadelphia university nexus learning. http://www.philau.edu/nexuslearning/

Philadelphia University. (2014). Hallmarks program for general education. http://www.philau.edu/hallmarks/

Philadelphia University. College of Design Engineering and Commerce. (2013). DEC outcomes. Unpublished manuscript.

Philadelphia University. Strategic Plan. (2009). Philadelphia university is developing the model for professional education in the 21st century. http://www.philau.edu/strategicinitiatives/

Chapter 11
The Master of Design in Strategic Foresight and Innovation (SFI) at OCAD University

Lenore Richards and Nabil Harfoush

11.1 Program at a Glance

The Master of Design in Strategic Foresight and Innovation (SFI) is creating a new kind of designer:

- A strategist who sees the world from a human perspective and rethinks what is possible;
- An innovator who can imagine, plan and develop a better world.

Recognizing the increasing importance and need for new thinking skills to positively impact society, enhance business success, and manage complex organizational change, OCAD University launched this innovative program in 2009 to address these issues and opportunities. Originally launched as a part-time program, the Master of Design in Strategic Foresight and Innovation attracts mid-career professionals from various backgrounds and professions, who continue to work professionally and immediately apply their newly acquired skills in the workplace. With the addition of a full-time option in 2013, the program has caught the attention of a global audience and attracts applicants from around the world. Students from Mexico, Brazil, India, Costa Rica and Nigeria significantly enhance the diversity of the student body.

The program focuses on the application of foresight, business and design innovation methods to develop solutions that are strategic, transformative and sustainable—economically, socially, and environmentally—and that address human needs. It interweaves design methods with social science, systemic design, futures thinking and business design with the aim of providing the skills and knowledge to better identify critical issues, frame problems differently and develop innovative strategies, solutions and implementation plans. Through holistic

L. Richards • N. Harfoush (✉)
SFI Program, OCAD University, 100 McCaul St., Toronto, ON, Canada, M5T 1W1
e-mail: nharfoush@faculty.ocadu.ca

B. Banerjee, S. Ceri (eds.), *Creating Innovation Leaders*,
Understanding Innovation, DOI 10.1007/978-3-319-20520-5_11

thinking in a co-creative environment, the artist, designer, entrepreneur, social scientist and engineer learn how to develop together the skills required for innovation leadership.

The SFI program attracts applicants with a diverse range of undergraduate and graduate degrees. It is a course-intensive degree program, which integrates a number of disciplines and extends the knowledge base in innovation methodologies. The program culminates in a Major Research Project (MRP) – a significant exploration of a topic of the student's choosing and employing much of the knowledge and skills developed in the program. The MRP topic areas are as diverse as the students themselves, and include healthcare, education, media and communication, sustainability, social innovation and foresight methods.

To address uneven knowledge and skills across our diverse cohorts and to enhance our curricular offerings, a week of bespoke workshops is organized each fall and winter semester. Included are workshops on business fundamentals, introduction to design, modeling collaboration, staging a presentation, and reverse archaeology (creating artifacts from the future).

The SFI program is the largest graduate program at OCAD University with an ongoing enrolment of over 100 students. Every fall, 22 students are admitted in each of our full- and part-time programs.

11.2 Short History

OCAD University, Canada's first art and design institution, was established in 1876. It is the third largest of the approximately 40 professional art and design universities in North America. OCAD U became degree-granting in 2002, and immediately embarked on the development of graduate programs and scholarly research activities. The Faculty of Design established its first research centre in 2005, defining a mandate in strategic foresight and innovation. An obvious next step was to develop a graduate program that would be inextricably linked to the research activities in order to engage with, learn about and articulate problems, principles and values common to design and innovation.

A steering committee was struck and work began in 2007. The final program proposal was submitted to the accrediting body in the winter of 2009 with an expected launch in September 2009. The program was promoted, subject to approval, in January 2009. It immediately attracted a diverse group of highly-accomplished professionals passionate about making change. The 2008 economic recession in North America underlined even more for these applicants the need to adapt to change and to be strategic in how one went about it.

External reviewers, key to the approval process, were skeptical about this non-traditional program that this very young university was proposing. They questioned our faculty qualifications, the value of the knowledge mix we were proposing, and the employability of graduates. Arrangements were made for them to meet the exceptional applicants who were eager to join this groundbreaking program. In the end, it was the students who convinced the reviewers of the potential of our program.

Approval was granted in late June of that year. Students were officially admitted and curriculum development began immediately. Students were advised that the program would be developed in collaboration with them and that it was understood that the reality of teaching such a program would be very different from the paper proposal. By its very nature, the program would be constantly evolving. In applying the iterative design process to the program itself, there will be continuous shifting as we explore more effective ways to teach and respond to changes in the external environment as well as to the expectations of the students.

The students in this first cohort were enthusiastic, and provided excellent and constructive feedback. Adjustments were made where possible: Smaller changes over the course of the first year, but more significant adjustments for the subsequent year. Feedback is continuously collected through faculty and student discussion as well as annual student surveys. A faculty retreat is held annually to explore in more depth issues that have arisen that year.

The research lab, the Strategic Innovation Lab (sLab), is strongly linked to the success of the program. sLab develops innovative solutions by applying design thinking, business intelligence and strategic foresight to envision alternative futures. sLab is a hub for collaborative relationships between OCAD U and government, industry, not-for-profit and community groups, combining advanced methods of academic research, project-based consulting and participatory engagement. The research projects provide important opportunities to test our methods – critical for informing curriculum development as well as providing the students with opportunities to apply their learning. Faculty-led research groups, including the Strongly Sustainable Business Model Group (SSBMG), SystemCity, Envision Health, Media Futures and Design Emergence Media Organization (DEMO), have emerged as foci of research within sLab. A number of other OCAD U research labs, including SuperOrdinary Lab and Situation Lab, again complement and provide thought leadership for the SFI program. A speaker series as well as seminars titled "Explorations" attract the OCAD U community and relevant external community audiences to interact, explore and discuss new thinking with notable thought leaders.

11.3 Educational Mission

The challenges we face today are increasingly complex in nature and in scale. These 'wicked' problems cannot be solved by any single entity, and there is a need for unprecedented levels of collaboration across disciplines. The SFI program was established to prepare innovation leaders who can navigate complex change and provide leadership in this rapidly changing environment.

As the first academic program in the world at the intersection of foresight and design, SFI can claim a place at the cutting edge of pedagogy and foresight practice. Combining design thinking and business thinking with futures thinking is key to improving strategic intervention and innovation. Leveraging the broad and deep pool of knowledge in design thinking at OCAD U, our aim is to become the pre-eminent learning environment where design and foresight meet.

Specifically, the objective of the SFI program is to graduate professionals, who will design creative processes, strategies, and implementation plans for collectives that want to make transformational change. The graduates are strategists and innovators, who work in the private, public and not-for-profit sectors.

Students learn to think creatively and holistically – exploring, challenging and finding meaning in order to reframe and guide both present and future actions. The program curriculum focuses on breadth, connecting the skills and expertise of the students to form effective groupings that can address complex problems.

The opportunities for the graduates include:

- Enhancing their professional skills and increasing their impact on their current organizations
- Career advancement to leadership positions in current or new organizations
- Bridging the transition to new but related professions
- Entrepreneurship in general and social entrepreneurship in particular

The following sample of research questions from the students' Major Research Projects illustrates how students apply their new skills to very diverse topics:

- How might we redesign the response of primary care to better address the social determinants of health of patients?
- Given changes in function, ownership and technology, how might designers create domestic objects that would have relevance over the next ten years?
- How can developing countries plan for housing market transformation that addresses both national goals and international climate change targets?
- How can we disrupt people's eating patterns and make nutrition a higher priority for all stakeholders in Toronto?
- As the internet enables audiences to play a more significant role in storytelling, how might the relationship between content producers and their audiences change?

We believe that our success is a result of the following priorities:

Application to 'Meta' Problems

The program focuses on the development of knowledge and skills to create transformational change. Research methods, design thinking and foresight tools are applied to 'meta' problems in a variety of sectors. Topics of exploration include health futures, strongly sustainable business models, media futures, urban systems, visualizing emergence, and education futures.

Student and Faculty Diversity
The students are typically mid-career professionals from the public, private and not-for-profit sectors and represent a broad range of disciplines, experience and cultural backgrounds. Our faculty are both practitioners and scholars and consist of futurists, designers, systems experts, social scientists, ethnographers, engineers, business consultants and strategists.

Transdisciplinarity
Multi-disciplinarity is key to addressing complex issues. The curriculum is focused on breadth, and draws on business, social science, design and foresight. In developing a curriculum that crosses many disciplinary boundaries, and through fostering exceptional faculty and student collaboration, a holism or transdisciplinarity is achieved.

Collaboration and Co-creation
An exceptional level of collaboration is needed between faculty, and between students. The course curriculum is intertwined, most project work is team-based, and assignments are often coordinated between concurrent classes. The depth and breadth of knowledge represented by both students and faculty is critical to making a co-creation environment possible.

Foresight is a differentiator for the SFI program. Our graduates can significantly enhance an organization's success by looking further into the future at changes that

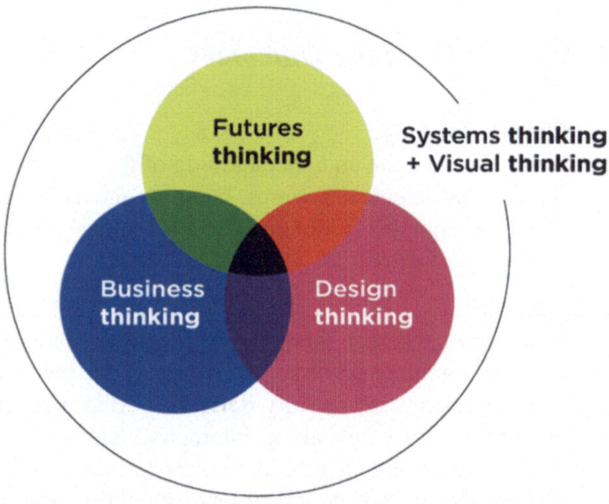

Fig. 11.1 Strategic Innovation Model. Our innovation process includes: problem finding (foresight), problem framing (strategy), and problem solving (design). Great emphasis is placed on unearthing or finding the real problem and then considering the different ways it might be framed before any action is taken to solve it. Living with ambiguity through extensive research, analysis and synthesis phases is critical for discovering the real issues that underlie a problem and will ensure more intelligent and responsive solutions

may be coming. This allows an organization to make smarter decisions today to 'future proof' itself – a very critical advantage in a rapidly changing environment.

The following is the most apt definition we have found for foresight:

> Foresight is a systematic, participatory, future intelligence gathering and medium-to-long term vision-building process aimed at present-day decisions and mobilizing joint actions.
> EC funded FOREN Project

11.3.1 Strategic Innovation Model

Our strategic innovation model illustrates the integration of design, business, and futures thinking through systems thinking (Fig. 11.1). This integration allows our students to move through an iterative design thinking process, understand the business context to ensure true sustainability and develop deeper insights into the challenges a sector or organization might be facing through futures thinking. Systems thinking and mapping locates these complex challenges in a larger system and makes clear the patterns and interconnectedness of the issues; and visual thinking ensures more effective communication of complex data. At the centre of it all is the focus on the human factor. Key to the success of any innovation is how well it addresses human needs, motivations and behaviors .

11.3.2 SFI Program Learning Outcomes

Depth and Breadth of Knowledge
Students will be able to demonstrate an understanding of the key theories, concepts and vocabularies in the social sciences, systems, business and design relevant to the definition and framing of complex problems, which in turn provides a critical foundation for applied project work.

Research and Scholarship
Students employ a variety of research methods from classic scholarship to design research methods, including interviewing, observation and participatory workshops. They must complete and successfully defend an MRP that takes the form of a written paper supported by visual or interactive elements to effectively communicate their outcomes. Students articulate a research question, develop and apply appropriate methodologies, and demonstrate original thinking in the analysis, synthesis and insights developed that address the research question.

Application of Knowledge
Students are continuously directed and encouraged to apply skills and methods learned for addressing organizational, societal or business challenges. They learn to synthesize and analyze the information collected to develop deep insights that can lead to creative strategies and solutions.

Professional Capacity/Autonomy
Working with industry ambassadors, students develop strategic proposals, actionable strategies and business models and methods to improve the sustainability of organizations. Students also demonstrate competence in facilitating participatory innovation processes.

Communication Skills
Students prepare concise, coherent and effective communications materials appropriate for their project outcomes, including written documents and visual materials such as graphs, maps, charts, photographs and drawings, as well as interactive experiences. They provide clear verbal presentations to articulate their project outcomes to their peers and industry mentors.

11.4 The Community

Students
The greatest strength of the SFI program is the deeply knowledgeable, experienced and passionate students. They are highly diverse; but what coheres the group is their strong desire to make a difference, to impact people's lives in a meaningful way. They are a very creative, optimistic and driven group of individuals, who enthusiastically and intensely explore some of society's most critical challenges.

As has already been mentioned, we believe that the diversity of the students is critical to developing deep insights into key challenges. Prime SFI candidates are mid-career professionals representing diversity in subject matter expertise, age, gender, worldview and cultural background. Their professional experience is in the public, private and not-for-profit sectors and their disciplinary backgrounds include the sciences, the social sciences, the humanities, computer science, engineering, design, art, digital media, strategy, marketing finance, economics, law and journalism.

We very consciously 'design' the diversity of each of our cohorts. We create a matrix of the students' attributes and carefully consider the makeup of each cohort as we make our admissions decisions. Academic achievement is only one of the many criteria used to predict success in SFI – we also consider length of professional practice, level of responsibility, diversity of professional experience, personal maturity, life experience and the students' motivations for joining the program. Emphasis is placed on exceptional interpersonal skills and professional accomplishment.

Faculty
Our faculty are a highly diverse group as well, due to the multi-disciplinary curriculum. There are both practitioners and scholars and they consist of futurists, designers, systems experts, social scientists, ethnographers, engineers, business consultants, and strategists. Due to the curricular imperative to apply new knowledge in real-world contexts, it is critical that the majority of the faculty be

practitioners. As a result, many are part-time, which is anomalous in a traditional university setting, although less so in professional programs.

As with the students, the faculty are as deeply committed to transformative change and believe in the program's potential to create innovation leaders. Most of the faculty were involved in developing the program initially, and have been engaged since 2009 in its continuous evolution. In addition to encouraging student collaboration in the classroom, courses are often intertwined and assignments are shared between concurrent courses. This high level of course collaboration significantly enhances the learning, creating connections between course content that might not otherwise occur.

11.5 Methodology

The SFI program is broad in scope, teaching skills and methods that can apply to a variety of problems in diverse sectors. The curriculum draws on specific theories and concepts from a number of disciplines as well as research, foresight and problem-solving methods that together create a body of knowledge that can be used to effectively problem find, problem frame and problem solve.

Co-creation is the underlying principle in all of our activities. The classroom environment is highly interactive. Students are typically given readings before class so that class time can be used for critical discussion. Faculty and students together engage in information sharing and ideation where the faculty act more as guides than as masters.

Application of knowledge and methods is a priority in a program focusing on change making. All learned theory is in the context of relevance to application. Case studies are examined to enhance real-world understanding and industry ambassadors are invited to participate in many of the courses. They provide a real- world context and subject matter expertise for course project work as well as valuable mentoring. With OCAD U's homegrown expertise in 'studio-based' education, many of our classes provide opportunities for in-class project work and individual coaching.

Teaching highly diverse classes does create some challenges, as there is an inevitable variance in knowledge and skills across our student cohorts. In order for our multi-disciplinary collective to become a more holistic trans-disciplinary collective, we mount targeted workshops in a 'workshop week' each semester. We also use these workshop weeks to address curriculum gaps that we have identified through the ongoing refinement and evolution of the program. Our workshop weeks are highly adaptive and serve as an excellent tool for immediate curricular intervention.

Opportunities for students to apply their knowledge are provided outside of the classroom as well. The Strategic Innovation Lab (sLab) as well as other OCAD U research labs hire the students as research assistants for scholarly and/or contract research projects. In addition, there are several faculty-led research groups where interested students are invited to participate. A recent collaboration with a business school teamed students from both institutions to explore innovative health care related ideas in the context of our aging population.

Transformative change-making skills in the SFI program are driven by the development of the following competencies:

Collaboration and Facilitation

Collaboration skills are critical for success in the SFI program, but more importantly for participating in and leading innovation processes. Representatives from a range of disciplines must work together to provide the requisite skills and knowledge when addressing complex issues. Our classroom diversity provides a perfect 'test bed' for this scenario.

Similarly, engaging key stakeholders is critical in the innovation process, both for the information and insights they can provide and for including and empowering them in the process. If stakeholders participate in the innovation process, they will be advocates rather than resisters of change. Facilitation skills are essential for providing this leadership.

Lectures, seminars and workshops that teach teamwork success, effective communication and negotiation skills and facilitation methods are provided throughout the program so that students can apply these skills most immediately to course project work and ultimately to real-world problems.

These skills include:

- Understanding different problem-solving styles and preferences and identifying them in a collective or team. We have consistently used Min Basadur's Creative Problem Solving Profile to identify each SFI student's particular problem-solving approach. This deepens the student's understanding of their own and their peers' particular skills and supports the development of teams with complementary skill sets.
- Identifying team member's skills, motivations and roles early in the project.
- Establishing processes and rules for decisionmaking in teams in the early stages of a project.
- Adapting communication styles to the particular composition of the team.
- Learning how to express dissenting views within a team and positioning such views as part of the divergent phase of the design process.
- Developing conflict resolution and mediation skills for use within and outside the team.
- Developing facilitation skills for small and large groups.
- Creating a new (non-hierarchical) team performance-based definition of leadership.

In order to create opportunities to practice these skills, we are more often assigning students to teams for course projects instead of allowing them to self-select. The intent is to create conditions similar to those of the real world, where professionals cannot always select their consortium team partners. More immediately, this allows us to 'design' the teams to ensure complementary skills and knowledge.

Tensions often arise between the team members when they are subjected to tight deadlines. We endeavour to be proactive when this occurs, and to use such situations as opportunities for students to practice leadership and collaboration skills. In this context we are experimenting with various self- and peer-evaluation methods that reward such practices.

Sensemaking and Futuring

Our Human Factors, Research Methods and Systems courses and our Foresight Studio build knowledge and skills to make sense of complex information and develop the insights necessary to properly find and frame problems. Fundamental human factors and systems theory and concepts are learned in the context of their applicability to social and business innovation. Ethnographic and participatory research methods ensure that people's needs are consistently at the heart of all research investigations.

In Human Factors students develop skills in practical research, critical thinking, synthesis and writing. In the Business and Design Thinking course the same students complete a cross-team project applying human factors and design thinking to an innovation challenge.

In our Innovation Research Methods class, students study and conduct research techniques for human-centred innovation, with strong emphasis on selecting and adapting appropriate methods for different problem types. A group research project investigates a social or innovation research question employing multiple methods.

We have a uniquely developed practice in our Systems course, where student teams select complex social system problems for which they conduct human-centred research and build system maps representing their research and design proposals. Visualization of complex data is critical in all project work and is particularly evident in the GIGAmaps, where the students create and visually communicate a narrative that provides a snapshot of the complex problem understanding and suggests opportunities for interventions.

Classic foresight techniques are studied and applied to challenges entitled 'The Futures of X' where topics such as healthcare, human communication and financial services are explored. Environmental scanning reveals trends and drivers of change that provide the critical uncertainties on which to build future scenarios. The culminating project in the Foresight Studio is the 'Time Machine' project, where students bring their futures to life. Time machines press design thinking, prototyping and transmedia storytelling into service as teams materially and performatively immerse visitors in future scenarios, whose logic they have spent the semester developing. The Time Machine deepens the understanding of how a scenario holds together, and with what strategic implications. It is the insights, implications and strategies that can be drawn from these possible futures that are so critical in helping organizations thrive in changing environments.

Ideation and Co-creation

In our introduction to the SFI program, the Business and Design Thinking course, design and business techniques applied to an identified problem space demonstrate the importance of design thinking for business success. 'Hands on' project experience is developed through a business simulation and the creation of an innovation design solution. Students are introduced to key design thinking methods, including problem definition, information gathering and analysis, idea exploration, idea testing/prototyping, and evaluation and selection.

Scenario development and the Time Machine project in the Foresight Studio call for significant ingenuity, holistic thinking and resourcefulness as students imagine future worlds five, ten or twenty years hence.

Strategy, Business Modeling and Action Planning
To meet our program objective of building skills to *make* change, the final three courses of the program emphasize actionable outcomes in the form of strategies, business models and implementation planning.

In the Strategy class, emphasis is placed on understanding the increasing need for and inherent challenges in developing coherent strategic solutions that drive effective organizations. Students learn to reveal the purpose and power of a strategy, to lead strategy development within an organization and to utilize purpose as a sustainable competitive advantage. The course collaborates with the Foresight Studio, sharing topics and student teams. Students work with the insights and implications from the Foresight Studio to define the strategy challenge, conduct analysis and develop final strategies.

The diffusion of new products, services and technologies requires new business models. The Business Model Innovation course offers the tools, methods and practices to analyze current business models, explore and evaluate new ones, and analyze the impact of regulation and legislation on them.

Collaborating with the Business Model Innovation course, Leading Innovation, provides skill development in leading and implementing innovation. Key concepts include types of innovation, styles of leadership, frameworks for change, how to position innovation, and how to overcome barriers to change. Student teams are assigned a real client organization and tasked with solving a business challenge using their combined learning from the two courses. They present their recommended solution in one integrated report to their client.

11.5.1 SFI Program Structure

The SFI program is comprised of 45 credits, 30 of which are required courses. There is a three-credit elective where students choose to take an elective course, complete an independent study project or participate in an internship. The program culminates in a 12 credit Major Research Project (MRP).

Part-Time Option Structure
The part-time option is a seven-semester program. The required courses take place one day per week for four semesters. Three semesters are dedicated to self-directed study in the form of an elective and the major research project (Table 11.1).

Table 11.1 Part-time Option Structure

Semester 1 Fall	Semester 2 Winter	Semester 3 Summer	Semester 4 Fall	Semester 5 Winter	Semester 6 Summer	Semester 7 Fall
Business and Design Thinking 3 cr	Understanding Systems 3 cr	Elective, Internship, Indep't Study 3cr	Strategy Dev't 3 cr	Business Model Innovation 3 cr	Major Project Proposal 3 cr	Major Research Project 9 cr
The Human Factor 3 cr	Foresight Studio 6 cr		Innovation Research Methods 6 cr	Leading Innovation 3 cr		

Table 11.2 Full-time Option Structure

Semester 1 Fall	Semester 2 Winter	Semester 3 Summer	Semester 4 Fall
Business and Design Thinking 3 cr	Innovation Research Methods 6 cr	Leading Innovation 3 cr	Major Project 9 cr
The Human Factor 3 cr	Strategy Development 3 cr	Business Model Innovation 3 cr	
Foresight Studio 1 3 cr	Foresight Studio 2 3 cr	Major Project Proposal 3 cr	
Systems Fundamentals 1.5 cr	Social Systems 1.5 cr	Elective, Internship, Independent Study 3cr	

Full-Time Option Structure

The full-time option is a four-semester program where two-and-a-half semesters are devoted to courses and one-and-a-half semesters to self-directed study – the elective course and major research project (Table 11.2).

11.6 Lessons Learned and Future Evolution

The program administration has been diligent in eliciting feedback from students through face-to-face discussion as well as through regular surveys, to ensure that it is meeting their expectations. This has provided the information needed for us to shift emphasis in the curriculum, adjust course sequencing and eliminate and/or add courses.

Focus on Application

The message we have consistently heard from the students is that they want to develop skills to *make* change. The focus must be on application, and any theory included in courses needs to be presented in the context of how it informs application. This has incited a good deal of healthy debate between the scholars and practitioners on the faculty, and we constantly strive to meet the students' wishes while maintaining an appropriate level of scholarly learning and work.

Teamwork

Student collaboration is one of the program's greatest strengths and is also its biggest challenge. Team selection, effective teamwork in a non-hierarchical environment, and differing motivation or contribution of team-mates are ongoing challenges. We have described some of the measures taken to address these challenges under "Collaboration" in the previous section. We learned that it is important for the faculty team to actively monitor and identify at an early stage the tensions emerging in student teams, and proactively intervene. We are considering using more frequent self and peer evaluation throughout a project as well as including team performance in the evaluation of course project work.

Professional Identity/Labeling

In a trans-disciplinary program, employers (and students) have difficulty articulating their skill set in an employment market still defined by vertical disciplines. In the first years of the program this resulted in student (and faculty) anxiety about the kinds of jobs graduates might pursue. The integration of SFI students in sLab research projects and the ambassador-supported class project work showcased the students' skills to clients, who began to worry less about assigning them a standard label. Today we have already an established track record of student achievement and employment in various sectors. As the reputation of the graduates spreads, this issue has all but disappeared.

The SFI program is about transformative change – but is also transformative for the students personally. We observed that many of them can struggle midway through the program, redefining their goals and professional identity. As mentioned above, this causes some anxiety regarding future professional direction, but tends to resolve itself toward the end of the program as students engage in their MRP, which many consider the stepping stone to a new career.

Program Growth

Growing an innovative program such as SFI is particularly challenging due to the difficulty in finding faculty, who truly understand the program and have, in addition to the academic and professional expertise required, the personal qualities essential to become part of the SFI faculty team. Beyond the challenges of creating suitable salary levels for part-time practicing faculty from a range of disciplines outside of the traditional art and design domains, there is the challenge of finding faculty passionate about creating new models of education; faculty who are exceptionally collaborative; and faculty willing to invest the time to engage with team colleagues and mentor and coach students outside of the classroom. Adding a full-time option was very challenging in this regard. Faculty who appeared eminently qualified to teach in the program struggled to meet the student and program demands, and in a number of cases simply did not meet expectations. Issues were resolved quickly, such as providing additional mentors to the faculty and bringing in guest speakers to address learning gaps.

The SFI program is young; the curriculum has been actively evolving since its launch in 2009, and in 2013 a full-time option was added. The program has grown from an initial cohort of 22 students to a total enrolment of over 100. There is no immediate need to grow in size, but for reasons of reputation and credibility, there is an interest in carefully considering the following options:

- partnering with other institutions either locally or internationally
- creating a low residency model that would attract part-time students from greater distances
- offering intensive certificate programs
- organizing student and faculty exchange with similar programs

In addition, providing more options or streams within the current program would allow targeted study in areas such as public policy or business innovation. Adding cohorts to our current programs would support the creation of streams.

Administrative Challenges

Traditional educational structures don't easily support new models of education nor the students they attract, and can be slow to respond to challenges that arise. With OCAD U, we are exploring new methods and arenas for recruitment and promotion, and new approaches and policies for nontraditional research projects, faculty appointments and admissions and registration processes. As an interim measure, we often develop 'workarounds' to accommodate student, faculty and operational needs. While there have certainly been growing pains, it is important to reflect and acknowledge what has been accomplished since 2009 as the SFI program and OCAD U together learn what is needed for this ground-breaking program to thrive.

Acknowledgement Special Acknowledgement to SFI Faculty and Program Co-developers: Jeremy Bowes, Stuart Candy, Stephen Davies, Peter Jones, Helen Kerr, Kate Sellen, Suzanne Stein, Greg Van Alstyne.

Chapter 12
Mission D, an Interdisciplinary Innovation and Venture Program at Tongji University

Lou Yongqi, Fan Fei, and M.A. Jin

12.1 Program at a Glance

The mission D program provides interdisciplinary "design-driven" innovation and entrepreneurship education to students at Tongji University, including its international and exchange students. The D in the name of the program refers to Design, not only classical skill-based design, but a way of creative thinking and acting. Here, "Design" is regarded as a common creative asset of human being.

Two institutions at Tongji University run the Mission D Program collaboratively: the Sino-Finnish Centre and the College of Design and Innovation. The mission D is a Minor Program, opened to students at different levels, including Bachelor, Master and PhD. The idea is to create an alternative educational opportunity for those who want to integrate and to apply knowledge and skills through "design thinking", to solve problems in different contexts. Together with the deep, operational and disciplinary knowledge and skills in a specific professional domain, these two capabilities are the basic elements of the so-called "T-shaped people". Mission D aims to cultivate T-shaped people who are "deep problem solvers in their home discipline but also capable of interacting with and understanding specialists from a wide range of disciplines and functional areas".

L. Yongqi (✉)
College of Design and Innovation, Tongji University, No. 281 Fuxin Road, Yangpu District, Shanghai, China

Sino Finnish Centre, Tongji University, Shanghai, China
e-mail: lou.yongqi@gmail.com

F. Fei
Sino Finnish Centre, Tongji University, Shanghai, China

M.A. Jin
College of Design and Innovation, Tongji University, Shanghai, Shanghai, P.R. China

B. Banerjee, S. Ceri (eds.), *Creating Innovation Leaders*,
Understanding Innovation, DOI 10.1007/978-3-319-20520-5_12

The curriculum is structured in three modules: (1) Inspiration & Introduction (2) Knowledge & Skills and (3) Integrative Project Experience. The Pilot edition is from February 2014 to June 2015, one-and-a-half years (three semesters) where the first semester focuses on courses and the second & third semesters focus on projects. In the pilot program, after completing 30 Tongji Credits (60 ECTS), students are awarded the Minor Certificate. If students cannot fulfill all the Minor requirements but complete a part of courses, they can get the relevant transcripts.

Furthermore, the curriculum for innovation and entrepreneurship education is integrated into a broader and well-designed ecosystem. It builds the entrepreneurial curriculum with an interdisciplinary, open and collaborative approach as well as a problem-based methodology involving universities, companies and public and private customers and consumers.

As a result, the program not only encourages collaboration between students across a variety of disciplines such as architecture, design, business management, linguistics, psychology, engineering, and biology, but also closely relates to emerging social and economic real-world challenges.

12.2 Short History

In 2009, Tongji University separated the design school from its College of Architecture and Urban Planning, and named it the College of Design and Innovation (D&I). D&I inherited a competitive faculty, a strong reputation, and a durable legacy; yet it is more than the sum of these. The transition from department to college signals growing capacity and need for new forms of design education. More importantly, D&I has a vision. The vision is to create a strategic paradigm that can bridge education, research and social impact around a design agenda. D&I aims to become a centre for design innovation and knowledge creation in the emerging industrial transformation of China. Design is also regarded as an important asset to generate knowledge production reform at Tongji University.

In order to extend design from “design doing” to “design thinking”, which can have more impact on the economy and society, Tongji needs an interdisciplinary platform at the university level. It becomes strategic for Tongji to work together with an international partner who shares the same vision and mission. Aalto University in Finland is that kind of ideal partner. On 9th January 2010, the second day of Aalto University’s official opening, Aalto University and Tongji University established a strategic partnership to found the Sino-Finnish Centre (SFC) together.

The Sino-Finnish Centre was planned as a strategic cooperative platform for innovation, sustainability and new business development based in Shanghai. The mission of the SFC is to actively contribute to the building of creative and sustainable societies in the twenty-first century through high-quality research, education and innovation. It provides students with new and novel learning opportunities and possibilities for deepening their knowledge to face real-world challenges in a global context. The SFC initiates R&D projects to support strategic cooperation between countries, universities and companies.

In 2013, the two universities co-established the interdisciplinary Minor program called the Tongji Innovation and Venture Program (TJIV), which is the sister project of the Aalto Ventures Program (AVP), an educational concept implemented at Aalto University. From 2014, it was called mission D. Capital D refers to design at thinking and strategy levels. Through the collaboration between the Sino-Finnish Centre, the College of Design and Innovation, Tongji Venture Valley, industrial partners, and so on, the program offers interdisciplinary courses and learning experience, which closely relate to real-world challenges. It has been the most popular minor program at Tongji University.

12.3 Educational Mission[1]

Mission D aims to cultivate T-shaped interdisciplinary "design thinking" talents, who will discover their capabilities in being creative, and put their innovative ideas and concepts into practice and start ventures(Leonard-Barton, 1995). As Tim Brown maintains: "[T-shaped people] have a principal skill that describes the vertical leg of the T—they're mechanical engineers or industrial designers. But they are so empathetic that they can branch out into other skills, such as anthropology, and do them as well. They are able to explore insights from many different perspectives and recognize patterns of behavior that point to a universal human need (Brown, 2005)."

To educate T-shaped (Figure 12.1) people with competitive advantages, the connection between vertical and horizontal knowledge and capabilities becomes crucial. The T-shape conception, however, addresses little on how the vertical part and the horizontal part connect. The SFC was established exactly for facing this challenge.

Mission D sheds light on various solutions to this problem. They focus either on fostering horizontal capability that can integrate multiple lines of vertical knowledge and skills (horizontal capability with depth), or on cultivating vertical

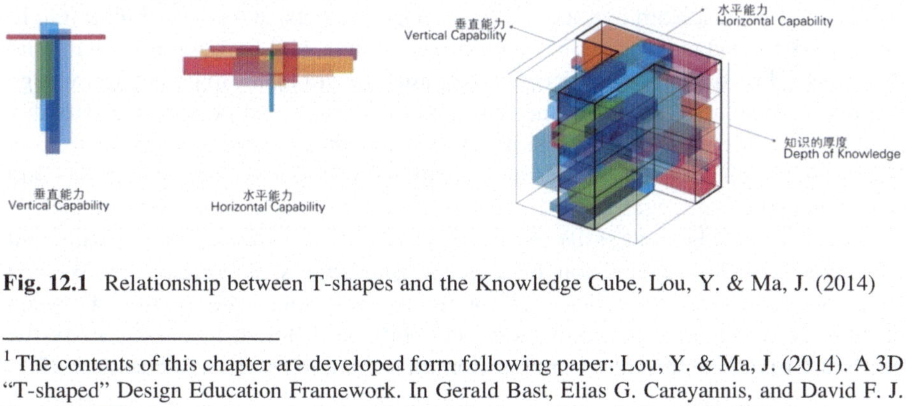

Fig. 12.1 Relationship between T-shapes and the Knowledge Cube, Lou, Y. & Ma, J. (2014)

[1] The contents of this chapter are developed form following paper: Lou, Y. & Ma, J. (2014). A 3D "T-shaped" Design Education Framework. In Gerald Bast, Elias G. Carayannis, and David F. J. Campbell (Eds.), Arts, Research, Innovation and Society. Cham, Switzerland: Springer.

knowledge and skills that also contain the vision to expand horizontally (vertical capability with broadness). Students need to obtain broad vision and the ability to integrate know-how from different vertical knowledge. To achieve this, students are encouraged to collaborate in interdisciplinary teams and work on real-world challenges. Collaboration with industries is always the best way. The students can apply and develop their know-how and creativity in a real context while receiving the technical support from the industry experts; the industries can also use the opportunities to explore the unexpected possibilities, from products to strategies of the companies, inspired by the young minds.

Mission D encourages collaboration between students across a variety of disciplines. One of the approaches is to recruit students from various backgrounds, for example, business and management, linguistics, psychology, engineering, and biology. The flow of knowledge is a common trend for most of the modern disciplines. Design, however, is the arena where this idea is highly appreciated and practically exercised. It relates with the interdisciplinary nature of design and the growing complexity of subject matters that design addresses.

The Mission D program provides an education-scape where the students can develop. It constitutes the educational environment, including curriculum, experience, practice, research, and life. A school is the organizer of a variety of knowledge, capabilities, spaces, and resources, and cultivates an ecosystem for each individual learner. Within such an ecosystem, every student may develop a unique growing trajectory. This conception is in concert with Dewey's observation of education:

> We never educate directly, but indirectly by means of the environment. Whether we permit chance environments to do the work, or whether we design environments for the purpose makes a great difference (Dewey, [1916] 2012).

In this sense, we design environments for our students. An even bolder attempt of mission D will be to further blur the boundaries: not only enabling students from different disciplines and backgrounds to collaborate together, but encouraging undergraduate, Master, and PhD students to co-study real-world challenges under the umbrella of international, university-business collaboration. Situated in the inter-cultural, interdisciplinary, and cross-education system environment, students will generate ideas, learn to grasp skills, manage projects, and conduct research. In this way, knowledge transfers between project teams, and between university and the society. Practice allows undergraduates and postgraduates to obtain knowledge from more than a single discipline with a sufficiently broad perspective (Hobday et al. 2011). We see this as a viable experiment that is able to respond to Don Norman's criticism on general design education—trying to encounter complex and comprehensive challenges from the real world but failing to do so due to a lack of necessary knowledge and skills (Norman, 2010). In its essence, the experimental educational mode we envision here is, as Buchanan's elaboration on Wicked Problems indicates, using design thinking to break the rigid barriers between disciplines, and integrating knowledge and skills at different levels to satisfy the needs and to realize the values arising from this changing world (Buchanan, 1992).

In conclusion, Mission D emphasizes on the following objectives:

- Design Thinking to integrate the knowledge of creation, business and engineering
- Core business skills such as international marketing, operations, strategy, finance, logistics, etc.
- Approaches to international product/service design and prototyping/piloting
- Leadership, networking, and boundary-spanning in cross-cultural teams
- Applied skills such as negotiation, teamwork, public speaking, and pitching ideas
- Working in cross-cultural teams to develop a plan for commercializing a new product/service for an existing organization or a new venture in China, Europe, and internationally

When designing the pilot Minor curriculum for mission D, the working team set the features of the program as follows:

- Keywords: Design, Innovation & Entrepreneurship
- International students' background: Chinese & -International students in one class
- Interdisciplinary: Design, Management, Engineering, Technology, Science, Humanity, Social Science, etc.
- Accessible and flexibility: open to undergraduates, graduates as well as Ph.D. candidates; individual course or whole program, encouraging inclusive mindset
- Pedagogical thinking: emphasizing Lectures + Projects (real life), using the Living Labs concept and the design thinking methodology

12.4 The Community

The Sino-Finnish Centre is generating interdisciplinary innovation by combining a variety of resources. It acts as an innovation hub: open, international and interdisciplinary, which invites cooperation among universities, companies, organizations and experts. It encourages the creation of human-centered and interdisciplinary approaches to the development of products and services. Based on a creative and interactive environment, it inspires learning by means of dialogue among students, professors, researchers, industry and business to experiment, prototype and interact for purposes of theory and practice. SFC has grown beyond Finnish-Chinese cooperation through the engagement of leading universities and industries from Europe, the US, Asia and Latin America in its activities.

Openness and sharing are two of the key characteristics of the SFC. The benefit of building an open platform is to harness the cumulative effort of several kinds of intelligence, resources, knowledge and ideas. The first meaning of openness is to test the boundaries of different disciplines, as we mentioned before. The second meaning of openness refers to the physical space. The space where the SFC is located is called the Aalto-Tongji design factory (ATDF) (Figure 12.2), which is a

Fig. 12.2 ATDF Space and Activities, ©Sino-Finnish Centre

part of the Design Factory Global Network started by Aalto University and now distributed in Finland, China, Chile, Korea and Australia. It is a place to encourage intercultural and interdisciplinary collaboration through a creative environment. The layout and function of the space is open in that it is both diverse and flexible. Diverse spaces interact with programs of different functions, including personal research, group discussion, teamwork, presentation, socialization, entertainment and so on.

Experiential learning is described by David Kolb as "the process whereby knowledge is created through the transformation of experience". We strongly believe that the traditional one-way of knowledge transition is due for change. The SFC encourages students to learn in proactive ways rather than through passive acceptance. The space of the design factory was designed to be full of delight, relaxation and pleasure. The students improve their knowledge through learning from the different people that they encounter at the SFC. At the SFC, students always take the center stage; they are the major drivers of many courses, programs and events. Thanks to this open strategy, the SFC has already been the most attractive place for Tongji and international students. It is almost like a university version of United Nations.

As both Tongji and Aalto are the core members of CUMULUS, the International Association of Universities and Colleges of Art, Design and Media, the only global design school association, with more than two hundreds member institutions, the same space has also been endorsed by Cumulus as the first CUMULUS Lab in the world. For Cumulus, it is a platform for promoting and presenting Cumulus in local regions; it will help other members of Cumulus to develop their contacts with local Cumulus members, and to build activities with them. The Lab also encourages the local Cumulus members to strengthen their connection with the Cumulus Association. The collaboration with international organizations such as CUMULUS and DESIS also greatly enlarged the community of the SFC.

The Minor program is offered to Tongji students, both Chinese and international, as a flexible scheme. Students from different schools and programs could choose to either register for the whole certificate program or register for just individual courses. During the first year, there were 38 undergraduate, master and PhD students registered for the whole Minor certificate and altogether there are over 350 students per year registered for courses.

The professors and teachers of the Minor are from different places. Tongji and Aalto University are developing activities such as faculty exchange, student exchange, joint intensive courses and joint research projects. In the first round of 20 courses, Aalto professors will either teach or co-teach with the Tongji counterparts. This is based on pedagogical cooperation between the two universities starting in 2011. The faculty groups in the two universities have established mutual understanding and personal contacts so as to facilitate the teaching practice in the Minor. Apart from the one-third of the faculty from Tongji and one-third from Aalto, the program recruits guest lecturers and entrepreneurs from the industry. The SFC has invited around ten professors from different backgrounds as "In-house Professors" and also hosts visiting scholars and innovators as Innovation Supervisors. They are the most dynamic teaching resource of the SFC. Every year at the SFC, there are more than 60 open lectures given by our visiting guests, the name list includes Thomas Friedman, Don Norman, Tim Brown, Gunter Pauli, and Kenji Ekuan.

12.5 Methodology[2]

The Mission D program adapted the Living Labs approach as the key feature of the methodology. It is a systemic pedagogical approach. It bridges different stakeholders, including the best universities, industries, communities, governments, and entrepreneurs, to create a collaborative platform to support the interdisciplinary education, research, entrepreneurship, incubation and speed up, which can help to enable positive social and economic changes.

Living Labs Approach: A Potential Element in Entrepreneurship Education Many universities label their entrepreneurship education programs with the tag of innovation. However, how to combine these two terms, innovation and entrepreneurship, is open to discussion. To link these in practice in the implementation of the entrepreneurship education we introduce the Living Labs approach as a user-centered future-oriented methodology to design entrepreneurial curriculum, so as to combine the recent trends and opportunities in social construction of knowledge and in innovation into the discussion of the future entrepreneurship education.

[2] The contents of this chapter are developed from the following paper: Fan, F. etal. (2014). Using Living Labs Methodology to Design Entrepreneurship Education Program in China: The case of a Pilot Minor at Tongji University in collaboration with Aalto University. ENoLL OpenLivingLab Days 2014 Conference Proceedings. ISBN: 9789082102727. Belgium: ENoLL - European Network of Living Labs

Living Labs in the Development of Innovation Theory In the real-life environment with multiple stakeholders, the living lab transforms the traditional lab to a social innovation lab, which is situated in the real-life contexts for co-creating innovation with users. Living Labs has been recognized in the EU as one of the potential key approaches to developing the future economy, as a means for designing new services, new products, and new social structures.

The theoretical framework of this innovative model includes the definition, structure, values, methods, and other factors. These have been summarized in the five principles of Living Labs: sustainability, openness, authenticity, user-involved innovation and spontaneity, and for seven components: real-life environment, methods and tools, technology and network infrastructure, test user groups, Living Lab specialists, organizational management, and product investors. The user-involved innovation and real-life environment are the two major features of Living Lab innovation.

Living Labs-Based Entrepreneurship Curriculum
For applying the methodology of Living Labs to design the mission D curriculum, the following key components are proposed.

- **Real-life environment**
  The teaching activities shall not be confined to discussion and knowledge transfer from teachers to students within the classroom and incubator. It is necessary to facilitate the students' experience in real-life environment, to observe and experience the market with users. In the physical space, by connecting teaching space, office space, and startup space in an inclusive and easily accessible way, the students can come into contact with entrepreneurs and innovators a daily basis. Such real-life exposure could lower the psychological barriers and shorten the distance to the entrepreneurial community. Living Labs particularly emphasizes real-life scenarios, and such a layout helps us realize the learning by doing concept advocated by John Dewey. The university is no longer a walled garden or an ivory tower, but an open innovation environment and part of a bigger real-life societal context.
- **Methods and tools**
  Using Problem-Based or Project-Based Learning (PBL), teachers and students can work in interdisciplinary teams with their different expertise and skills. Connecting to one principle of Living Labs: user-involved innovation, one human-ed pedagogy: design thinking is proposed for entrepreneurship education. Design thinking is solution-oriented thinking for practical, creative resolution of problems or issues with the objective of better future results. Design thinking, which combines creation, business and engineering, or what we call DESIGN, is the most influential pattern adopted in the innovation education society. The D-School at Stanford University, Alta Scuola Politecnica (a joint effort of PoliMI and PoliTO) in Italy, and the Design Factory of Aalto University in Finland are a few successful endeavors based on this pattern. The PBL pedagogy, can guide students through certain phases of design thinking to solve specific problems and meet real needs. The study found that when applying design thinking in the teaching process, students could master the skills and

mindsets similar to those gained in business practices (Zupan and Nabergoj 2012). Some of these programs have even taken a step forward, i.e., to infuse entrepreneurship initiatives into design innovation education. Such an endeavor is innovative in every sense.

- **Technology and network infrastructure**
  Nowadays, the hardware equipment and technology support in Chinese universities are sufficient to provide the basic infrastructure for Living Labs-based entrepreneurship education. However, on the Internet infrastructure side, there is room for improvement. In addition, social networks in entrepreneurship and small business development are critical because business knowledge and skills can also be taught in social networks (Dana, 2001). Another reason is that entrepreneurship education not only includes explicit knowledge which can be taught but also tacit knowledge, such as experiences and insights which can only be learned by social interaction. It is necessary to design the technical network so as to facilitate knowledge transfer on social networks. Incidentally, new mobile Internet messaging and social networking services that would very well support the collaborative and social learning aspects as well as the rapid development of ad hoc purpose-driven people networks have become phenomenally popular in China. The primary example is Weixin (WeChat in the West) service by Tencent that provides a potential platform for use in education.
- **Multi-stakeholders**
  According to Living Labs theory, multi-stakeholders include test user groups, Living Lab experts and product investors. Entrepreneurship courses at colleges and universities in the ecosystem perspective also involve students, full-time and adjunct faculty, entrepreneurs, investors, universities administrators, government officials, companies, as well as Living Lab experts and other experts and researchers in innovation and entrepreneurship.
- **Organization and management**
  To efficiently run the entrepreneurship curriculum, the organization and management can no longer be a top-down hierarchical guidance. The management team is of importance in open innovation of the curriculum, but open innovation of the curriculum implicitly means that specifically the students are developing the curriculum.

Here we would like to introduce two of courses in our mission D minor Program. The Mobile Course, also call Tongji on Tracks, is one of the most famous. It is a course that inspires students to tap their potential to create social impact via bringing cultures together and cooperating on real challenges in practical life. It was designed to take place on modes of transportation such as trains, boats, buses, etc. The students are encouraged to use design thinking and an interdisciplinary approach to finish a project provided by the industry. Normally, the topic will be co-defined by the company and the SFC. The first Mobile course started on May 2011. Eighty Tongji Students from 30 different colleges and departments took a train from Shanghai to Helsinki. Before, during and after the trip, the students are required to finish projects provided by leading companies such as Haier, Kone, Volvo, among others. The students who successfully finish the course can get two credits from Tongji. The course provides a unique experience for the students and puts them into a complex scenario that will greatly challenge their capacity of innovation and collaboration. The combination of a defined innovation challenge and the changing environment, provide an ideal pedagogic setting. Such a kind of training can never be achieved from normal, disciplinary courses.

The second example is the Product Development Project (PDP). It is primarily aimed at students of engineering, design, and business who are interested in product development and are at the final stage of their studies. The course is becoming broader and master's students from other fields (e.g. medicine) are also starting to find their place in PDP projects. Most of the problems are given and sponsored by companies searching for innovative cooperation with the next generation of product developers. In the beginning, much attention is directed to team formation to create highly motivated, interdisciplinary teams. A project typically includes the phases of planning, concept creation, decision making and detailed computer aided development. Finally, manufacturing, assembling and testing have also proved to be valuable learning experiences. In 2013, for instance, collaborating with Panasonic, the SFC hosted a successful PDP project named "InnovAIR." Participating students, from different countries and five disciplines, proposed a creative solution to the PM 2.5 air pollution problem. Aside from proposing design concepts, they accomplished the technological solutions and planned a highly feasible business model. Figure 12.3 captures a few moments while the students were making the functioning prototype.

Fig. 12.3 On-site scenes of the InnovAIR project, the PDP course offered by SFC. ©Sino-Finnish Centre

12.6 Lessons Learned and Future Evolution

The assignments of the courses typically end with a presentation. Self-esteem and positive emotions in difficult situations develop remarkably through the oral pitching. After enough practice the students are mature to defend their opinions and plans in public, which changes their psychological threshold.

Among the courses, one assignment is to recognize an opportunity, create a business model, evaluate the model, and pitch the proposal. In some user cases, the students are the real-life users (e.g. second-hand furniture sellers and buyers, or electric bicycle riders). For business-to-business case authenticity, several operating companies are invited to presents real-life cases with necessary business information. The student assignment groups are self-organized, which quickly makes it clear how important it is for an entrepreneur to choose the right trusted partners. Also, the need for multi-disciplinary skills become evident when drafting the business model, including market studies, marketing, sales, finance, human resources, innovation management, production, and many other aspects of real-life operations needed in any company. The assignment as a whole is a tough road to go. However, most students survive and complete the course with high self-confidence, motivated spirit, and enriched practical skills in many areas, including the skills to collaborate or even lead a team.

In the students' feedback questionnaire after one semester of study, we solicited qualitative data. The 38 registered students are expecting the following from mission D (Table 12.1).

While being asked about their rewards after one semester, students report the following outcomes (Table 12.2).

The Mission D Program is carrying on during the 2014–2015 academic year with a new partnership with Stanford ME310 and Global Design Innovation Course. And

Table 12.1 Feedback questionnaire from students

How to think more open-mindedly and more real practice opportunity.
Product marketing strategy combined with product design.
Confidence management. Leadership Design program.
Different disciplines with innovation information.
New way of thinking is a new kind of mind-set.
Diversity, group work, how to sell ideas.
Maybe some knowledge of physics but in an innovative way.
Negotiation skills.
About design and innovation.
Time management.
How to make investment.
Solid ability of co-working.
Idea generation.
Business design/model.

Table 12.2 Report from students about the reward

The atmosphere, and the chance to work with international student.
The way we work as a team to find solutions and generate ideas.
Get to know people from different education background, multi-discipline
Foreign friends/Multicultural.
Air of open learning and open innovation.
Getting new ideas & ways of thinking.
Mind-set of creation or innovation.
More organized way of innovative thinking.
English improvement.
Broadened my horizon/widened my views/opening my minds.
Flexibility.

the joint team will continue to explore the educational offering for the students in the diversified group with new challenges from the economy, society and technology.

12.7 Conclusions

The challenges of this era are no longer defined in disciplines. We need a new and robust research tradition to investigate and explore the issues involved in models of education as well as models of practice. We need a theoretical foundation to support new ways of research and education. We need an innovative and inclusive social ecosystem that will enable the application of new knowledge and methods to generate social and economic changes.

It is also important to strengthen the cultural dialogue between the western world and the non-Western world on design and innovation. It is time to rethink the value of non-Western culture for sustainability and for providing alternative measures of quality and performance that move beyond systems that require high energy consumption and high resource consumption. Developing economies in nations are possible test beds in which Design can drive dramatic paradigm shifts towards a new economy and a new society. The dynamic social context in these countries presents a great opportunity for the application and development of Design Thinking (Lou, 2013). The result of implementing DESIGN should not be another kind of monolithic single solution for all problems. The SFC and D&I encourage the same rich diversity, complexity, and inclusiveness that is the key to sustainability in nature.

Through Mission D, we begin to experiment in a proactive way, connecting and sharing experiments through critical and reflective practice, improving design tools and striving for generalizations that can apply in a variety of contexts. The scale of impact should then be measured not by individual size, but by its ability to generalize to a wide range of situations. These kinds of networked, small interventions can have significant systemic impact, helping us to generate a new restorative economy and a new society. The Mission D Program at Tongji University is an attempt exactly towards that direction.

References

Brown, T. (2005). Strategy by Design. Retrieved September 28, 2013, from Fast Company, website: http://www.fastcompany.com/52795/strategy-design

Buchanan R (1992) Wicked Problems in Design Thinking. Design Issues 8(2):5–21

Dana LP (2001) The education and training of entrepreneurs in Asia. Education + Training 43 (8/9):405–416

Dewey, J. ([1916] 2012). Democracy and Education: An Introduction to the Philosophy of Education. Indo-European Publishing. (Originally published: New York: Macmillan, 1916.)

Eriksson M., Niitamo VP, & Kulkki S (2005) State-of-the-art in utilizing Living Labs approach to user-centric ICT innovation-a European approach. Lulea: Center for Distance-spanning Technology. Lulea University of Technology Sweden: Lulea. Online at: http://www.cdt.ltu.se/main.php/SOA_LivingLabs.pdf.

Zupan B, Nabergoj AS (2012) Developing Design Thinking Skills in Entrepreneurship Education. Leading Through Design. Boston: Design Management Institute

Fan, F. et al. (2014). Using Living Labs Methodology to Design Entrepreneurship Education Program in China: The case of a Pilot Minor at Tongji University in collaboration with Aalto University. ENoLL OpenLivingLab Days 2014 Conference Proceedings. Belgium

Hobday M, Boddington A, Grantham A (2011) An Innovation Perspective on Design: Part 1. Design Issues 27(4):6–15

Leonard-Barton, D. (1995). Wellsprings of Knowledge: Building and Sustaining the Sources of Innovation. Boston: Harvard Business School Press

Lou YQ (2013) Design Harvests, an Acupunctural Design Approach towards Sustainability. Mistra Urban Futures, Gothenburg, Sweden

Lou, Y. & Ma, J. (2014). A 3D "T-shaped" Design Education Framework. In Gerald Bast, Elias G. Carayannis, and David F. J. Campbell (Eds.), Arts, Research, Innovation and Society. Cham, Switzerland: Springer.

Norman, D. (2010). Why Design Education Must Change. Retrieved September 28, 2013, from Core77, website: http://www.core77.com/blog/columns/why_design_education_must_change_17993.asp

Chapter 13
Alta Scuola Politecnica: Innovation, Multi-disciplinarity and Passion

Marco Ajmone Marsan, Stefano Ceri, Roberto Verganti, and Roberto Zanino

13.1 The Program at a Glance

Alta Scuola Politecnica (www.asp-poli.it, (ASP 2009)) was founded in 2004 by the Politecnico di Milano and Politecnico di Torino. It annually selects 150 talented students solely on the basis of merit, from among the applicants to the Master of Science in Engineering, Architecture and Design at the two universities. The resulting community is made up of very motivated and qualified students with a passion for innovation. These students follow an additional track, completely in English, based on ad hoc courses and the development of multidisciplinary projects; the ASP program administrates 30 credits (CFU) to be achieved in parallel to the regular 120 credits of the Master program, 15 from courses and 15 from projects; thus, it consists of an addition to the Master program of about one-fourth of its total workload. The mission of ASP is to provide society with high-profile graduates combining in-depth disciplinary knowledge with interdisciplinary, horizontal skills that are needed to work in a truly multidisciplinary environment.

13.1.1 Courses

ASP courses focus upon interdisciplinary issues and innovation; they are held in the form of six intensive, residential schools, for a total of about 70 hours of classes per year; all travelling and lodging expenses are covered by ASP. The teaching style is

M.A. Marsan • R. Zanino
Politecnico di Torino, Corso Duca degli Abruzzi 24, Torino, Italy

S. Ceri (✉) • R. Verganti
Politecnico di Milano, P.za L. Da Vinci 32, 20133 Milano, Italia
e-mail: stefano.ceri@polimi.it

B. Banerjee, S. Ceri (eds.), *Creating Innovation Leaders*,
Understanding Innovation, DOI 10.1007/978-3-319-20520-5_13

a mix of formal lectures, seminars, team work sessions and personal work; the emphasis is on learning instead of teaching, with strong interaction between lecturers and students. The current courses are:

- **Innovation and Society**, stimulating students' sensibility for the broader economic, social, cultural, ethical, juridical and political contexts of professional work.
- **Design Methods,** introducing the wide range of existing methods, from the descriptive models for analyzing design processes and behaviors, to the prescriptive tools that provide a structured and multi-disciplinary approach to design.
- **Management of Innovation,** providing the students with a comprehensive understanding of innovation from the point of view of both management and economics.
- **Complex Decision Making in the Public and the Private Sphere**, providing a theoretical framework and analytical instruments for managing decision-making processes.
- **Global Change and Sustainability**, introducing the problems of global change, policies that shape sustainability paths, and social commitment to economic and environmental sustainability.
- **The Dynamics of Creativity**, exploring how creativity and innovation can help us to face the ongoing situation; recently, while the world has been facing a global crisis, the course has taken as specific focus the promotion of "development beyond de-growth".

13.1.2 Projects

Students participate in multidisciplinary projects, focusing on real and relevant problems, proposed as collaborations between universities and external institutions (i.e., firms, the government or research institutions). Project teams consist of five or six students from different schools and are therefore intrinsically multidisciplinary. Each project team is assisted by tutors with multifaceted backgrounds, who are members of faculties of the two universities or of external institutions. Tutors advice students and periodically review their intermediate results. Teams are provided with funds that allow them to make study trips, attend conferences, and buy books and materials to help develop the project. The project plan and budget are defined by the team and approved by the tutors. The results of the projects are published in books that is issued yearly and can be downloaded from the ASP web site.

13.2 A Short History

While several excellent education initiatives are present in Italy, both in the basic sciences/humanities fields (e.g. Scuola Normale Superiore, www.sns.it/en/) and in the applied sciences/technical fields (e.g. Scuola Superiore Sant'Anna, www.santannaschool.eu), ASP is a school of excellence focused on multi-disciplinarity and innovation, unique at the national level; it is the result of a joint academic project involving two major technical universities, both with 150 years of history, and capitalizes on the strategic alliance between Milano and Torino, two productive cities of northern Italy connected in less than one hour by a high-speed train.

ASP was initiated by a grant of the Ministry of Education, which allocated globally about 11 M Euros in September 2004 to the two founding universities. The grant was awarded after a selection process open to the entire Italian university system, giving them a lot of freedom in the program implementation; the main constraint was to merge the faculties and students in the endeavor. After an intense activity of curricula design, the school was opened in the winter of 2004 with a first intake of 150 students. In 2014, ASP selected its XI student cycle, and has graduated so far over one thousand students (who receive a double degree from the two universities as well as the ASP diploma).

The ASP board is formed by eight professors from the two universities; decision making is concentrated within the board, which is appointed by the Rectors and by the Senates of the universities and has full autonomy on all the ASP administrative functions, such as students' admission, curricula organization (course and project selection), and conferring of the ASP diploma to students; the school has a lightweight organization with very limited dedicated personnel and uses services from the two universities (such as international admission or career services). The ASP director and vice-director from the two universities are members of the board, the director is appointed for a three-year term.

In the beginning, ASP had to face skepticism from those of the two faculties who considered such a broad-spectrum initiative as being inevitably superficial, as well risking of being confused with a business school. This situation has significantly improved over the years, thanks to the increasing involvement of the faculty in both courses and projects (so far about one hundred faculty members per cycle have been involved as tutors of projects or as teachers or tutors of ASP courses). By adopting this strategy, ASP makes the best use of the spectrum of disciplinary competences present in the two universities.

13.3 The Educational Mission

In short, the vision of ASP is to support bright students in designing smart solutions to complex problems, while coping with dynamic environments which require:

- cognitive capacity and ability to learn fast;
- capacity to manage organizational processes;

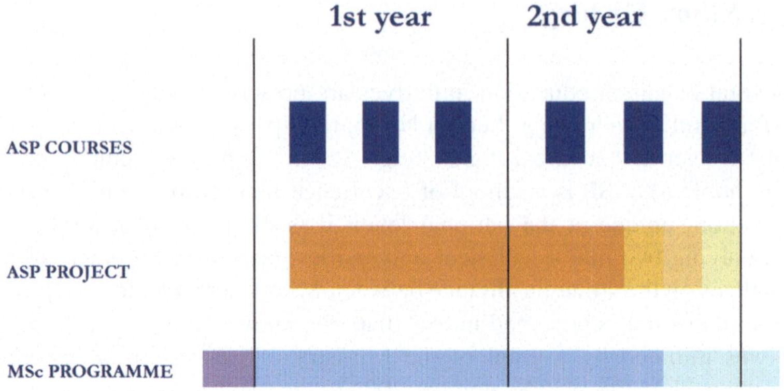

Fig. 13.1 ASP courses and projects along the two-year M. Sc. program

- awareness of the interests and the interactions that take place within the specific context;
- flexibility in adapting to cognitive processes used during the design process;
- talent for interpersonal relations.

This vision is implemented through a general pedagogical approach aiming at the following objectives:

- To set the foundations of a critical approach to the main facets of complex, interdisciplinary design;
- To explain and teach the implications of innovative design in terms of history, sustainability, socio-economic impact, and strategies for successful entrepreneurship;
- To complement and complete the role of M. Sc. studies by providing an all-round education.

Courses and projects engage the students during the two-year ASP cycle, from the beginning of the M. Sc. second semester to the M. Sc. graduation at the end of the second year, as shown in Figure 13.1. Projects run throughout the ASP program, and courses are scattered throughout the two-year cycle of ASP, with a progression that has been designed so that they can provide foundational knowledge while projects are progressing.

Technically, courses are organized as follows: the content, organization and program are defined by a coordinator, appointed by the ASP board after selection and after accurate tuning of the course content and methodology. Coordinators are world experts in the course domain; in some cases they are faculty members of the two universities but in other cases they come from other institutions. The coordinator, in turn, selects the speakers and the tutors of the course (typically three to four professors and four to five tutors, sometimes more) and organizes the scheduling, typically consisting of morning lectures and afternoon discussions among students in work groups led by tutors, followed by general discussion in a plenary session guided by lecturers and tutors.

Courses have a studio style, where lecturing is intertwined with group activities; most courses present a problem in the a form of a week-long challenge; students respond to the challenge by articulating solutions through stages; eventually, the solutions are compared and the best solutions are acknowledged. It is customary for students to spend long nights in preparation for presentation or in response to challenges. Each course is associated with required and recommended readings (in advance) and proposals/traces for paper preparation; students must deliver individual work after each course, in the form of two long papers (graded) and four short papers (pass/fail).

The ASP projects enable concrete experimental activity on broad and relevant themes and problems. Therefore, selected problems are system-level, interdisciplinary, and focused on problem-setting within a complex situation, where innovation (i.e. bridging technology with its applications) has a substantial role. Projects emphasize conceptual design and feasibility analysis, but in many cases they have been conducted up to prototyping or even implementation and have given rise to about ten spin-off companies so far. They are very different from master's theses, and require both disciplinary knowledge (coming from the master studies background of each student) and interdisciplinary knowledge (coming from the ASP courses).

Projects are proposed by teams of academic tutors and external institutions (i.e. companies, public bodies, NGOs) or by ASP investors; they are selected by the ASP board after a formalized process (call/revision/acceptance). Each project has an academic tutor who is responsible of providing the required domain background, and an external tutor who is responsible of presenting the external institution's viewpoint in terms of needs, requirements, and objectives. In many cases the members of the ASP student team perform several site visits at the stakeholder's premises.

Student teams are built matching the students to the projects, based on their preferences and on the skills required, and subject to diversity constraints (w.r.t. nationality, Politecnico of enrollment and the Master's curriculum); each team consists of five or six students. Teams are given a budget covering expenses for travel and limited experimentation; the ASP board interacts with the students for project management activities. Figure 13.2 shows the various players of ASP projects.

During the development of projects, students are trained in abductive thinking as a way to develop their innovation and leadership capabilities. Different from inductive and deductive thinking, it implies the capability to imagine new situations that are still not in place, which is necessary in a world that keeps changing. Abductive thinking can only be developed through project-based education beyond specific disciplines; as suggested by Roger Martin (Martin 2009), leadership in the current socio-economic context requires abductive thinking. This is the reason why our program gives significant importance (50 % of the curriculum) to projects. Also, immersion in a multidisciplinary environment enables the development of skills to become a conductive leaders in increasingly multicultural environments.

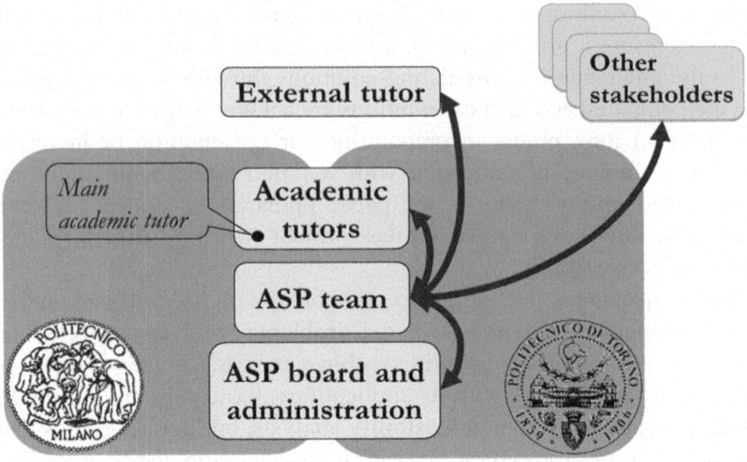

Fig. 13.2 Players of ASP projects

On the basis of our experience, projects can be clustered into two groups:

- Technology- & research-driven (*technology push* → moving from research results to applications); projects in this group address a technological innovation scenario characterized by a narrower multidisciplinarity, where the field has already been explored; they typically deliver feasibility studies and prototypes.
- Design-driven (*demand pull* → starting from a complex problem and trying to find solutions); they are characterized by a broad multidisciplinarity, and focus on innovation scenarios that are mostly unexplored from technical, structural and functional points of view; they deliver a final study that explores the problem situation, compares possible solutions, and selects the most promising and innovative options.

Projects can be further broadly categorized as related to sustainability, social inclusion and health/quality of life, hi-tech and net economy, and energy and mobility management, as shown in Fig. 13.3.

Altogether, Politecnico di Milano and Politecnico di Torino award each year about 25 % of the engineering and 40 % of the architecture and design Bachelor titles in Italy. ASP benefits from the merging of these three cultures (engineering, architecture and design): systemic innovation, which is produced by ASP projects, typically results from the strong technological background of the engineering students, from the awareness of the context framework of the architecture students, and from the attention to user needs of the design students (see Fig. 13.4b). If one considers instead the contribution that each student population can make to solve systemic problems, we note that engineering students typically provide a solid logical and mathematical approach, architecture students provide a perspective towards composition, and designers provide visionary approaches (see Fig. 13.4a).

The cultural program of ASP reflects the 2009–2010 criteria published by ABET, Inc. (the former Accreditation Board for Engineering and Technology), prescribing that engineering programs should have for their students a series of

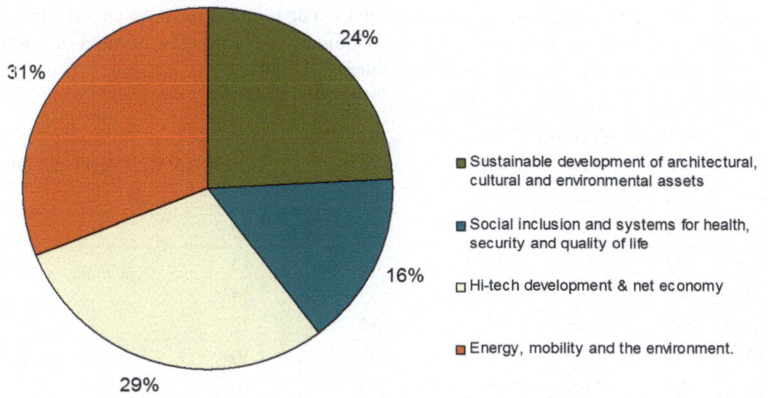

Fig. 13.3 Distribution of ASP projects

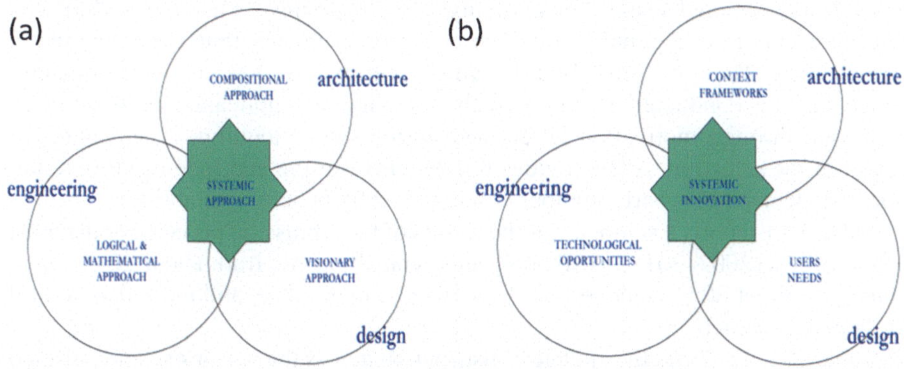

Fig. 13.4 ASP merges the cultures of engineering, architecture, and design

outcomes, including the ability to function on multidisciplinary teams, the understanding of professional and ethical responsibility, the ability to communicate effectively, a broad education necessary to understand the impact of engineering solutions in global, economic, environmental and societal contexts and a knowledge of contemporary issues. Other analogous European documents, such as the British UK-SPEC, the French CTI References et Orientations, the German ASIIN Requirements and Procedural Principles, the international EUR-ACE Framework Standards (all available online) contain similar requirements.

13.4 The Community

ASP students are selected through a differential selection process which considers intakes from the two universities, from other Italian universities, and from international master's applicants. Applicants must satisfy tight selection criteria (e.g. BSc

Table 13.1 ASP student population, years 2007–2013. The target is to admit exactly 150 students at each cycle, with a ratio 60 % to 40 % between Milano and Torino. The number of applications reflects small differences in the admission procedures

	APPLICATIONS		ADMITTED STUDENTS		
CYCLE	MI	TO	MI	TO	INTERNATIONAL STUDENTS
IV (2007)	174	110	83	57	26
V (2008)	251	230	90	60	39
VI (2009)	293	200	90	59	35
VII (2010)	240	255	90	59	34
VIII (2011)	219	146	90	57	27
IX (2012)	379	210	90	60	30
X (2013)	**291**	**166**	**90**	**59**	**20**

grades in top 5 % at the time of application and English knowledge as assessed through international tests), but normally, after applying such criteria, ASP can select students from a population which is five times greater than the actual target for selection. Thus, the ASP Board invites applicants to a selection colloquium, which can be conducted remotely with international applicants; motivation as expressed through motivation letters and during the colloquium is an important aspect of the selection. ASP provides full benefits (tuition and lodging) to international students, their yearly number is influenced by benefit availability.

Table 13.1 reports a summary of the results of the admission process for the most recent cycles (2007–2013). ASP applicants typically come from about 20 different countries, about 60 % achieve their B.Sc. degree cum laude, and more than a third of them are women.

Table 13.2 describes the student's distribution across the various schools offered by the two universities; it also shows that students from architecture and design are about one third of the ASP population, reflecting the students' distribution in the two universities. In general, the ASP's target is to distribute students more or less proportionally to their distribution within the schools of the two universities.

One important aspect of ASP is the creation of a network of people who share the same values and passions. Along this dimension, students decided to create the ASP Alumni Association (http://alumni.asp-poli.it/), for upholding the community feeling between students, who decided to keep in touch, not only through the natural friendship born during the ASP experience, but also by creating a stable net of contacts and relationships and to form an organization that will be the point of reference for all the students and graduates of the present and future cycles of the ASP. Currently the association has about 800 members working in more than 20 countries; their placement is shown in Fig. 13.5.

The alumni association organizes yearly conferences together with field trips and networking events; a timeline is shown in Fig. 13.6.

During 2012, the Alumni ASP association started a mentoring project, which consists of creating a one-to-one partnership between a mentor (who is an ASP alumnus) and a mentee (who is an ASP student), who meet several times during the development of the ASP cycle for sharing knowledge, skills and perspectives and to

Table 13.2 ASP students' distribution across the various schools of the two universities

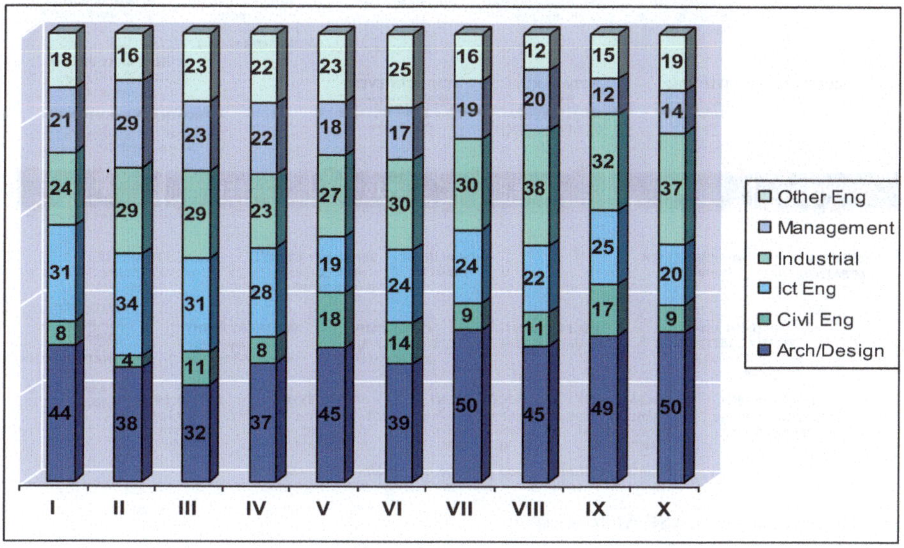

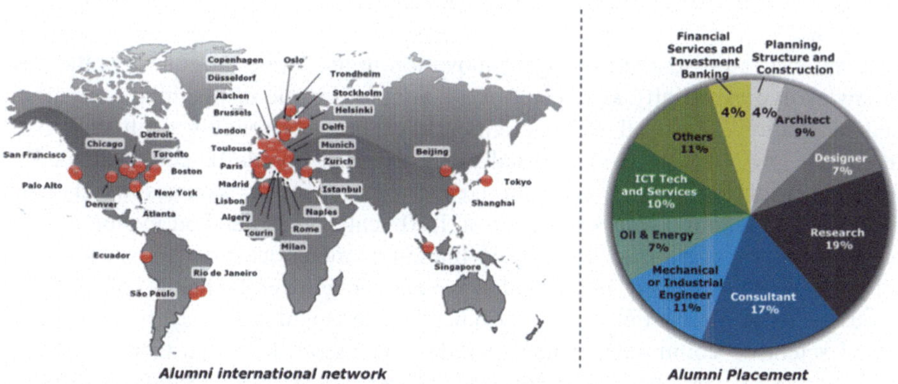

Fig. 13.5 ASP alumni distribution in the world and in the various placement sectors

foster the student's personal and professional growth as well as his or her transition from the university to the professional world. In general, ASP encourages contacts between ASP Alumni and enrolled students and is particularly proud of results that are autonomously produced by the Alumni Association.

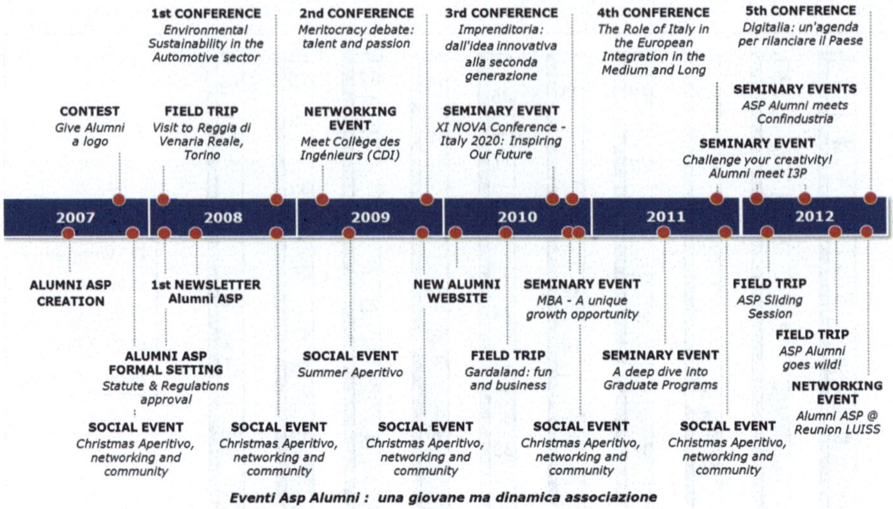

Fig. 13.6 Timeline of ASP Alumni events

13.5 Feedback and Future Evolution

Evaluation is an essential aspect of innovation in academic programs, as the programs evolve essentially as a result of a continuous cycle of planning, execution, and evaluation. Thus, all ASP activities have been thoroughly evaluated, both internally (by students) and externally (by external boards or consultants). For example, Figure 13.7 shows randomized evaluation of six courses which have been offered through four cycles, on a 1/10 scale; very detailed evaluations of each course are performed through on-line anonymous questionnaires. Similarly, all projects are evaluated; in particular, we encourage peer evaluation of students' contribution to project outcomes. Periodically, the Board is advised by an international scientific committee, which includes Professors R. C. Armstrong, (Cambridge, USA), M. Ferrari, (Houston, USA), E. Goles, (Santiago, Chile), K. Osterwalder, (New York, USA), and L.Yongqi (Shanghai, China).

In 2013, ASP commissioned an external evaluation, which was performed by Istituto di Ricerca Sociale (IRS) located in Bologna. A long report (in Italian) was based upon interviews with the main players of ASP (students, board, tutors, alumni, external companies, sponsors) and upon a critical assessment of ASP's first ten years of activity. The results presented by the report are quite positive in terms of students' placement: about 93 % of the students of the 7th cycle are employed, a larger fraction than comparative student's cohorts has achieved a PhD after the ASP diploma, most students recognize an advantage in their relationship with the employers thanks to the ASP "imprinting". At the same time, the report indicated some problems in the organization of the curriculum and in the ASP "brand" communication, both within the universities and towards external

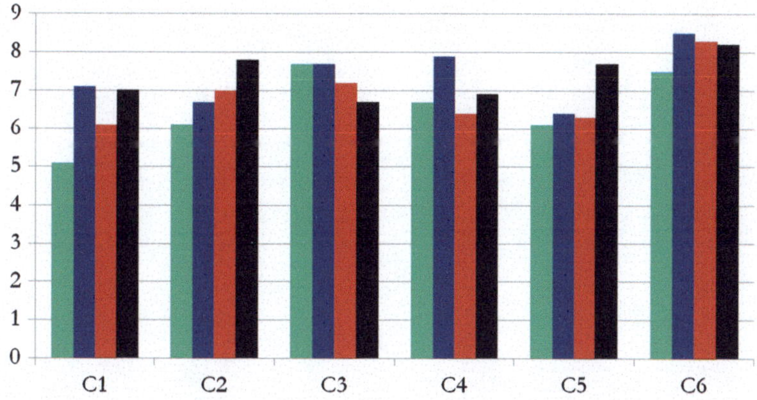

Fig. 13.7 Evaluations of ASP courses in a 1–10 scale, over four different cycles

organizations, thus pointing to the main directions in which ASP is called upon to evolve and to improve in the years to come.

As a result of this report and of a survey of the opinions of former students, after ten years of activity, ASP is revising its operations in order to adapt them to new external challenges and to start a new stage of growth. The ASP Board planned a set of new activities to be implemented in the three-year Board term:

- A new brand image has been designed, and is already reflected in the new ASP web site.
- A tighter coupling between the multidisciplinary project activities and industrial requirements has been implemented, by shortening the duration of projects, so as to provide industrial partners with results compatible with the life cycles of their products and innovation processes.
- A retargeting of course content is under way to exploit the vast body of competences of the two universities, and to carefully balance the impact of courses on multidisciplinary projects.
- Last but surely not least, a network of leading companies that collaborate with ASP is being built for a stronger alignment of the ASP curriculum with industrial expectations, as well as for the diffusion of the image and branding of ASP and for the creation of qualified employment opportunities for ASP students.

References

ASP (2010) Benedetto S, Bernelli Zazzera F , Bertola P, Cantamessa M, Ceri S, Ranci C, Spaziante A and Zanino R: Alta Scuola Politecnica: an ongoing experiment in the multidisciplinary education of top students towards innovation in engineering, architecture and design, European Journal of Engineering Education, 35: 6, 627–643.

Martin R (2009) The design of business: why design thinking is the next competitive advantage. Harvard Business School Publishing, Boston, MA

Chapter 14
The Paris d.school

Veronique Hillen

14.1 Program at a Glance

Paris d.school is a consortium of 5 schools (ENSAVT, ESIEE, UPEM, EIVP, ENPC), covering disciplines ranging from architecture and urban planning to all types of engineering, along with business and finance. It is supported by a French government grant of 4.1M€ distributed over 8 years through ANR (National Research Agency in France). The project for a French d.school was developed between 2007 and 2011, and was submitted in 2011 for the IDEFI tender organized by the French Ministry of Research and Education, with the objective of developing initiatives for excellence in innovative teaching in France.

It is of paramount importance for France to foster economic growth, either through strategic renewals of large companies or through the creation of start-ups. For this reason, future innovation leaders need to be trained to become intrapreneurs, in order to develop new economic activities in existing or newly created companies. Such profiles need to combine both expertise in a given field and the practice of innovation. The trans-discipline of design thinking, as developed and practiced in Silicon Valley, offers a solid source of inspiration for innovation in pedagogy development. This kind of culture of exploration needs nevertheless to be adapted to the French context and its specific areas of concern and excellence. Combining expertise and design thinking will require huge development efforts.

Paris d.school's mission is to become a demonstrator of future pedagogies in innovation through the trans-discipline of design thinking. After two years of operation, it already provides three levels of training, both for master-level students and professors: initiation workshops, intermediate courses, and expert programs (see figure 14.1).

V. Hillen (✉)
Dean, Paris Est d.school, 6 avenue Blaise Pascale, 77455 Champs sur Marnes cedex, France
e-mail: hillenv@dschool.fr

B. Banerjee, S. Ceri (eds.), *Creating Innovation Leaders*,
Understanding Innovation, DOI 10.1007/978-3-319-20520-5_14

Fig. 14.1 Paris Est d.school three-tiered courses

In 2014, it offered a dozen courses, corresponding to 82 ECTS (credits): 12 students in a full time Masters-level program in cooperation with Stanford (58 ECTS); 200 students distributed across half a dozen intermediate courses (from 2 to 9 ECTS); over 500 students initiated through workshops (0 ECTS); over 11,000 French citizens attended one of our lectures (either in auditorium or online). Students come from all disciplines and from different countries. A dozen professors from different disciplines have joined the d.school so far, and a further 60 have expressed an interest. In 2014/2015, 6 new courses are planned (Fig. 14.2).

Located in a new positive energy building since 2013, Paris d.school offers professors, students and partners a number of resources: a studio with different staging spaces, a prototyping room of over 600 square meters, a platform of pedagogical resources, prototyping and traveling budgets, pedagogical advice, a network of partners, peer-to-peer exchanges, and training programs for professors and coaches. Through its track record of projects, Paris-Est d.school aims to become a global benchmark for sustainable cities including the silver economy, energy efficient buildings and emerging economies.

14.2 Short History

The plan for the creation of a French d.school was developed through two previous phases, first with multidisciplinary programs at the national level, then with multidisciplinary multi-cultural programs at the international level:

From 2007 to 2009, three multidisciplinary multi-school courses were created and offered as part of the curriculum in the Industrial Engineering department at École des Ponts ParisTech (ENPC). During that period, high levels of tensions and failures were observed among the teams. A glue for linking disciplines was needed. Project outcomes consisted of pitches for ideas, with no evidence of an understanding of user needs and no impact. Alternative ways of teaching with a transdiscipline

	2012/2013	2013/2014	2014/2015	2015/2016	Moyens	ECTS	Nbre équipes projets	Nbre formateurs
Formation initiale - niveau master								
Initiation (sans ECTS)								
	0	450	0	450	Ateliers parallèles d'initiation « Tous Créatifs »		60	12
	0	160	0	230	Atelier pour le réseau international ME310 à Stanford		15	6
	180	0	50	50	Atelier «Culture Design Thinking » pour ESIEE			3
				50	Ateliers "entreprenariat"			4
Sous-total sans ECTS	**180**	**610**	**50**	**780**			**75**	
Intermédiaire (1,5 à 9 ECTS)								
	44	33	25	30	Innovacteurs, GI/ENPC	6	7	7
	44	44	29	30	ecobootcamp, GI/ENPC	3	6	3
	0	20	25	25	d.seniors, ENSAVT/ENPC/ESIEE/EIVP	5	4	3
	0	0	11	15	d.structure, ENSAVT/ENPC	6	4	3
	0	22	25	25	Création d'entreprises innovantes, SEGF/ENPC	3	7	3
	0	0	26	30	Gestion des projets d'innovation dans les services, UPEM	3	4	3
	0	40	32	30	Field trip (2 modules x 1,5 ECTS), ENPC	3	8	1
				12	d.city, à déterminer	9	3	3
					d.urbanspace, EIVP			
				8	d.building, à déterminer	9	2	3
Avancé (plus de 9 ECTS)								
		12		12	ME310 Design Innovation avec Stanford	58	3	12
Sous-total élèves avec ECTS	**88**	**171**	**173**	**217**	4 à 10 équipes projets/ cours (23 à 150 heures en présentiel)	**105**	**48**	
Sous-total nbre cours	2	6	8	10				
Sous-total nbre ECTS	9	78	29	105				
Nbre d'heures de cours								
Sous-total équipes-projets	13	32	40	48				
Formation continue								
Initiation (sans ECTS)								
	695	1800	1714	700	Nbre de participants à une conférence/workshop			
	6	11	14	10	Nbre de conférences données			
	4631	3382	9065	7000	Nbre téléchargements d'une de nos conférences			
	1	1	1	22	Nbre de conférences téléchargeables			
	10	21	120	200	Nbre de participants aux sessions de formation			
	4	5	6	10	Nbre de sessions de formations proposées aux formateurs			
Sous-total sans ECTS	**5336**	**5203**	**10899**	**7900**				
Intermédiaire (1,5 à 9 ECTS)								
	0	28	30	30	d.leaders (2 ECTS) / MBA Solvay ENPC	2	6	1
			10	10	Make or buy (1 ECTS) / MBA CNAM	1	2	1
Avancé (plus de 9 ECTS)								
			12	0	Master Systèmes Urbains Intégrés (12 ECTS)	12	4	2
Sous-total avec ECTS		**28**	**52**	40		**15**	**12**	
Total ANR		**5231**	**10951**	**7940**				

Dans le rapport ANR, le nombre d'heures assurées par des représentants du monde socio économique est calculé sur 10 à 20% du nombre d'heures enseignées (calculé lui-même par le ratio de 30 heures pour 1 ECTS).

Autres indicateurs								
						Nbre entr	Nbre visiteurs	
	15	40			Nombre de visites			

Fig. 14.2 Paris d.school three-tiered courses and respective ECTS

were therefore examined. The Center of Design Research at Stanford and design thinking as practiced in Silicon Valley became a fruitful source of inspiration.

From 2009 to 2012, ME310 Design Innovation was developed at ENPC in collaboration with Stanford and its international network, SUGAR (including the Aalto Design Factory and HPI). This full-time program in Paris has created successful conditions for ambitious innovative projects, with an ecosystem of about a hundred people worldwide, strong partnerships with large French companies, as well as the creation of a dedicated space and a strong community. Design thinking was a strong source of inspiration for the building of high-level competences in the practice of innovation with major real-world impact.

Since 2012, the d.school has been created following a successful bid to the French government, which meant a radical change in scale, in terms of space, and the number of classes and professors involved. The trans-discipline of design thinking has been adapted and developed to fit the French ecosystem and the nature of projects. Barriers have been observed, including a difference between the American culture of "let's do it" and the French one of "let's think about it". Design thinking has to be developed and adapted to the French context, notably the need to demonstrate its value with more explanations regarding "what, why and how", as well as to tackle ambitious challenges with technological implications.

The common challenges for each period were as follows: lack of understanding and support from the local ecosystem; the challenge of creating a start-up in the context of a public administration. The data supporting decisions were based on the discovery of design thinking in Silicon Valley as a source of inspiration and lessons from pedagogical action research with educational experiments in a French context.

14.3 Educational Mission

Paris d.school's educational mission is to prepare future innovation leaders by developing a culture of exploration (instead of a culture of exploitation). Paris Est d.school fosters both intrapreneurs (developers of new activities in an existing organization) and entrepreneurs (founders of new businesses), with either economic or socially-driven objectives (Fig. 14.3).

In France, economic priorities are linked with the ability of large companies to rejuvenate their businesses. It is the reason why it is of paramount importance to foster profiles which are strongly linked with the vision of innovation leadership explained in the previous chapters of this book: in the face of complex new problems, and especially to renew their activities and business models, large organizations need leaders who do not define themselves by their position of power in a hierarchical pyramid, but by the fact they out-perform "Business as

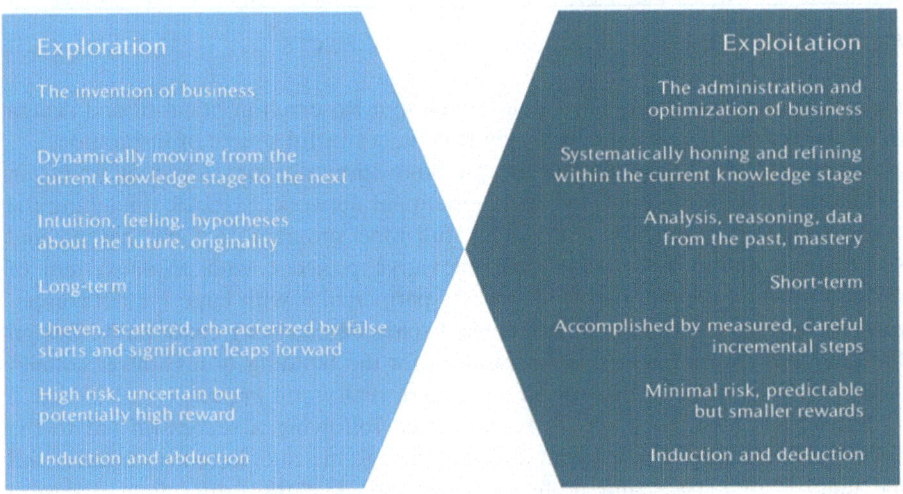

Fig. 14.3 Exploration v. exploitation (adapted from R. Martin, 2009)

Usual" processes and "conventional wisdom" in many ways (as explained in Chap. 1). The types of leaders we educate are able to create and leverage complex innovation ecosystems (as explained in Chap. 2). Paris Est d.school aims to educate innovation leaders with the right competencies and skills, not only to imagine alternative futures but also to drive breakthrough implementations with scaled impact (instead of managing the present with normative methods). Those leadership 2.0 profiles represent a fundamental frame-shift from the normative mental-model of a leader (as explained in Chap. 3): innovation leaders have transformative capacities when they amplify the impact and the innovation capacity of a system. In order to achieve this challenge, innovations leaders must be trained with a different mindset which requires a willingness to "see" possibilities and imagine scenarios that are different from those currently in place. It also requires a drive for impact combined with the ability to be comfortable with ambiguity. Innovation skills are also needed, in order to conceptualize, realize and amplify outcomes. The different levels of innovation leadership can be defined by the degree of complexity of challenges (challenge class) and the ultimate impact of outcomes (impact frame).

Paris d.school employs different pedagogical formats in order to acquire the different types of capabilities and skills in the expertise ladder of innovation leadership: basic workshops to disseminate the vision of innovation leadership, advanced courses to tackle class A problems and expert programs to tackle class B and C problems with the attributes listed in Chap. 3 (p. 78–84) as targeted pedagogical objectives.

The expected level of professional capabilities (in terms of knowledge acquisition, mindset, innovation skill and impact frame) depends on pedagogical formats, which are strongly linked with the number of teaching hours. All formats combine the development of a spirit of exploration, knowledge acquisition, as well as practical capabilities in design thinking in line with three types of exploration spaces (inspiration, ideation, implementation). Students acquire the ability to tackle ill-defined or wicked problems and create appropriate tangible solutions thanks to a creative confidence gained through the practice of methods and tools. Depending on course format, students acquire the spirit and skills to solve problems with different levels of ambition. They range from introducing a new dimension into an existing project to creating a new product/service offering, either as an intrapreneur (with the transformation of an organization) or as an entrepreneur (with the creation of a new company), through finding users and experiences for a given technology. All briefs reinvent experiences for users, either with objects, services, processes or the three combined (i.e. a system).

14.4 The Community

Paris d.school's ecosystem includes professors from its five academic partners, other faculty members, fellows, students, alumni and partners:

The **faculty** community encompasses Paris d.school's core team, professors and researchers from its 5 academic partners, as well as external fellows and independent experts.

(a) Paris d.school's permanent team has accumulated 7 years of experience in teaching design thinking pedagogy with direct responsibility for 2 to 6 classes a year representing up to 82 ECTS a year (with 12 to 44 students per class). After two years of operation, development and pedagogical experiments, the permanent team identified three levels of accreditation for faculty, which require different skills, different pedagogies and different levels of involvement: – initiation – intermediate – expert. Depending on the level of aspiration, training programs and coaching services are suggested. Faculty needs have been identified to define the role of Paris d.school as a resources provider and advisor.

(b) Paris d.school faculty members are offered a range of services, in order to transform their courses; as for February 2015, 12 professors from different disciplines (architecture, urbanism, city planning, languages, communication, industrial engineering) have joined and been initiated (workshops, travelling experience, co-development of new courses), and 60 from other disciplines have expressed an interest.

(c) External experts and fellows are recruited to help the development of new courses or to enrich existing courses. Depending on the nature of a project or the need for a course, they are recruited on the basis of their expertise and track record. A dozen experts were involved in the first two years in the following areas: network management, phenomenology, brainstorming, silver economy, social entrepreneurship in emerging countries, foresight for big cities, digital applications, ethnographic research, interaction design, landscape design, car design, software programming, furniture design, eco-packaging, sustainability, frugal design, product design. Experts deliver workshops, lectures, and coaching.

(d) A scientific committee seeks to nurture thinking and action in developing this pedagogy. Experts in design thinking worldwide advise Paris d.school and can participate in developing resources (an open-source video platform, recording lectures at d.school for broadcasting, the development of pedagogical supports, joint publications, recommendations...).

Students are mainly Masters-level graduates from academic partners of Paris d.school. Students can enrol in such courses, either directly by applying to the d.school for full-time programs or by taking courses from the existing curriculum; they come from all disciplines: all engineering fields (telecommunication, computer science, mechanical engineering, industrial engineering, telecommunications, data science, civil engineering...), urbanism, architecture, finance, economics, service design etc.

Paris d.school's **partners** provide projects and access to the field for our students to innovate. Each course has at least one partner and a maximum of 8 partners. Partners are structured into different categories, depending on the kind of partnership developed: 1- internal departments in each of the consortium's academic institutions, such as the library, a research lab or the human resources department; 2- local socio-economic agents, such as retirement homes, local organizations for human, environmental or animal rights, as well as economic development for the neglected in emerging countries, natural environments; 3- cities and public spaces; 4- start-ups; 5- large companies. They provide briefs and access to the field where students can carry out ethnographic research, tests with users and implementation. Depending on the nature of the projects and partners, they also provide resources such as financing.

Paris d.school benefits from the support of **alumni** through project suggestions, expertise, job offers, financing and event participation. Given historical developments, the community is mainly composed of alumni from ENPC's industrial engineering department and from ME310 Paris. It represents over 50 students a year from 2007 to 2012, and over 200 students since 2012, that is to say more than 700 alumni in all (80% French, 20% others, both in terms of nationality and location).

14.5 Methods for Teaching, Learning and Assessing Performance

14.5.1 Faculty Training

With regard to faculty, the range of services has switched from a three-phase training model (immersion, co-development, transfer) to a model of resource provision and an accreditation program. This change was based on feedback on the experience of the first two years. The first model was time and resource intensive, with low impact in terms of involvement, course transformation and skills development. The second model allows for greater responsibility, as faculty stay in control of their courses, as well as scalability, as faculty members can be at least tripled with the same level of resources. During the first two years of operation, the two models were trialled. The result was the development of a new course which enrolled 4 schools, 20 students and 4 professors (6 ECTS) and required at least 200 hours of the permanent teaching team, in comparison with another course (one professor, 4 departments in one school, 23 hours, 1.5 ECTS) which required 20 hours of development. For the same level of developed ECTS, the ratio is 1 to 2.5, which means the first model required more than the double the time!

Four kinds of resource are now developed and provided: access to space and potential funding, a platform of pedagogical resources, as well as training programs and advisory services for pedagogy development.

With regard to **space**, a studio has been structured with state-of-the-art pedagogical equipment and complete refurbishment with the help of professional interior designers, in order to foster the spirit of exploration and to provide activity spaces, such as a prototyping room, a library corner, project spaces, a kitchen, a collection of materials, a speaker's corner with room for up to 25 students, a living room for project reviews, a brainstorming room and zen spaces. The space also includes access to a 450-seat lecture hall with state-of-the-art lecture recording equipment and video broadcasting, 8 classrooms with appropriate equipment (such as smart boards, paperboards, modular seats and tables), a videoconference room and a 650m^2 prototyping space with lifting equipment.

A **pedagogical platform** is currently being developed to provide all kinds of teaching materials (video, syllabus, activities...) and will be regularly upgraded.

Training programs are offered to teachers, with **three levels of accreditation**: level 1 is called "d.ambassador" and recognizes teachers' ability to speak about design thinking and user-centered innovation at an expert level, as well as to conduct pedagogical activities in design thinking without the involvement of an actual partner; level 2 is called "d.coach" and acknowledges the ability of teachers to conduct a project with a local partner within the framework of their expertise and to coach student teams on such challenges (implementation of innovation which creates value for users in a local context); level 3 is called "d.leader" and recognizes the ability of teachers to conduct an ambitious project outside their own field of expertise, the ability to adapt design thinking to a specific area of practice (either user-centered, such as seniors, or field-centered, such as urbanism) and their stage directing capacities (i.e. the ability to set up ideal conditions for students on ambitious real-stake projects within a large ecosystem).

Advisory services will be also provided with three phases: course set-up (with syllabus, briefs, partnership, pedagogical objectives and activities, assignments and evaluation process, expert identification and selection if need be); course development (with expert involvement, real-time adjustments, pedagogical action research); course assessment (with self-reflexivity aligned with research activities, evaluation of student and partner feedback, project assessment, skills assessment).

14.5.2 Student Training

Students are offered three kinds of courses: initiation workshops, intermediate courses and expert programs. They have the possibility to learn either the entire process or part of it (such as one activity), the culture of design thinking (either adapted to their fields of expertise or outside their field), either as a compulsory or elective course, in a multidisciplinary or mono discipline team. Depending on their initial background, this offering differs and depends on faculty's involvement and teaching development in each institution. Global educational objectives and teaching methods are described in the section on "educational objectives". Examples are described above for each level of expertise, giving detailed explanations of the methods used for teaching, learning and assessing performance:

- **Initiation workshops** (0 to 1.5 ECTS, from half a day to two days):

  We run a dozen workshops and a dozen lectures a year, either for Masters-level students, professors and researchers, staff of each school, MBA students or company executives. Some lectures are on line. Most are free-of-charge. All of them are run by Paris Est d.school.

- **Intermediate courses** (3 to 9 ECTS, from 75 to 225 hours for students)

  We run half a dozen of these classes, and our target is to reach a dozen per year. These classes target Masters-level students from the consortium's schools. Students pay regular fees to their parent institution. Courses are provided by professors from Paris Est d.school's 5 schools. They are part of the Master's curriculum to the extent desired by professors and academic directors. The average number of participants is around 20 students. Each course tackles a specific group of users (seniors, fire-fighters, students. . .), local context (library, garden, urban furniture, eco efficient building, pedagogical equipment. . .) or area of expertise (architecture, urbanism, air quality, waste management, services in transportation, engineering of urban systems, cities, agriculture, app development. . .).

Example: "Ecobootcamp" for Masters-Level Students in Industrial Engineering

Upon completion of this course, participants should have gained creative confidence in all aspects of running a small-scale innovation project with the culture of design thinking. The course is a combination of knowledge acquisition at expert level through a series of lectures and pedagogical materials, development of cognitive skills through a number of exercises in controlled environments, as well as the development of practical skills in inspiration, ideation and implementation through a field project, with coaching and project reviews. Projects tackle real issues, with access to users for need-finding and testing, as well as the possibility of final implementation. For briefs, a sustainability dimension is added as an additional factor of complexity. Depending on the level of expectation, assessments include an exam, field activities, a final presentation and a project report. The impact and the degree of implementation, as well as the level of self-reflection, are key factors of performance.

- **Expert programs** (from 15 to 60 ECTS, full-time basis over one or two semesters). Paris d.school offers three kinds of expert programs: department or Masters project (15 to 25 ECTS) over a semester or a year, a full-time program (58 ECTS) over a year in collaboration with Stanford and its international network (*ME310 Design Innovation*), a full-time program (30 ECTS) over a semester in collaboration with local development associations in emerging countries (*Social Entrepreneurs*). Paris Est d.school's core team is heavily involved in staging the ideal conditions for students to foster innovation (partnerships, briefs, syllabus, teaching materials, lectures, project reviews, final presentations and exhibition, implementation). A dedicated space is allocated to project teams for the duration of the program. Briefs are structured to re-invent the product and service offering, and to create the conditions for its implementation in a given organization, which maybe a private sector, public sector or social sector entity.

Example: ME310 Design Innovation program
ME310 Paris has been in existence since 2009 and was initially led by the department of industrial engineering at ENPC. Masters-level students from all disciplines and countries work in close collaboration with leading companies within the framework of an international network led by Stanford (SUGAR). The objectives of ME310 Paris projects objectives are to re-invent a company's offering for specific users in a specific industry and to contribute to the implementation of at least a pilot in a real-world context. ME310 Design Innovation has been run since 1969 at Stanford.

14.6 Lessons Learned and Future Evolution

Feedback from **students** varies between workshops, expert programs and intermediate courses. All students from workshops and expert programs are 100% satisfied. Workshops always trigger a high level of curiosity and interest, to the point that some students wish to change their career direction ("I want to dig into design thinking for my MBA thesis, in order to find a job in that domain" or "I want to create my own consultancy company in that area"). Expert programs always trigger a lot of amazing feedback from all students, in terms of personal development and human transformation, unforgettable experience, and promising career prospects. The reasons for this satisfaction include strong initial motivation, the pride in implementing something real with strong impact, as well as a high level of recognition from partners and the community. The reactions of students on the intermediate courses are more qualified. Their satisfaction depends on their achievements and on responses from the ecosystem ("We do not see the point of quick-and-dirty prototypes and we get criticised by our mates"). Their achievements depend to a very great extent on their openness and commitment, team dynamics and skills, as well as the opportunities to implement their projects and their impact. The definition of the brief and faculty skills also play an important role. After their second class, 100% of these students in 2014 were satisfied and proud of what they had achieved.

As of February 2015, all our **partners** were fully satisfied with the implementation of the projects, which had exceeded their expectations ("I was not expecting such an implementation with a set of real equipment constructed in so little time"; "I had a dream and you have made it"; "I'd have never imagined that we would succeed in creating, industrializing and commercializing such great furniture for seniors in less than 9 months. We've never done it in our company").

More French **teachers** than expected are interested in this kind of pedagogy. Nevertheless, they quickly realize the level of commitment and effort required. Teachers lack the time needed to develop partnerships and to adapt design thinking to their field, which is the best way to demonstrate the value of design thinking.

In the future, Paris d.school has three main objectives: – to further develop the teaching skills of professors – to consolidate and develop its worldwide expertise in seniors, sustainable cities, public sector and emerging countries – to create a worldwide living lab to sustain projects and bridge the gap between courses and business development.

Chapter 15
Interdisciplinary Design in the College of Arts and Humanities at the University of Brighton

Anne Boddington, Jyri Kermik, and Tom Ainsworth

15.1 The Programme at a Glance

In the College of Arts and Humanities at the University of Brighton we have begun to explore emerging forms of design practice and the opportunities these present for the development of design as a core competency of twenty-first century life. Areas of particular interest are those in which design might be repositioned within university education, not primarily as a professional activity but equally as a vital and empowering portfolio of knowledge and skills that generate more confident and resilient graduates.

Design Futures, conceived as a conceptual model for an outward looking and adaptive design education, with a purpose to provide a shared space to drive innovation in design research and pedagogy, sits within the School of Art, Design and Media (part of the College of Arts and Humanities). Design Futures explores the tensions between design research and traditional and emerging professional practices of design. It also acts as an interdisciplinary catalyst to engage with students and researchers in other courses and academic programmes within the college and university, as well as with industry and innovation communities both locally and internationally.

This interdisciplinary development along with the culture of innovation championed by the College of Arts and Humanities enables students, academics, researchers, and professionals to connect more easily. It provides collaborative project opportunities for undergraduate and postgraduate study across disciplines including the arts, design, the humanities, medicine, pharmacy, computing and

A. Boddington (✉)
College of Arts and Humanities, University of Brighton, Brighton, UK
e-mail: A.Boddington@brighton.ac.uk

J. Kermik • T. Ainsworth
School of Art Design & Media, University of Brighton, 73 Embassy Court, Brighton BN1 2PY, UK

B. Banerjee, S. Ceri (eds.), *Creating Innovation Leaders*,
Understanding Innovation, DOI 10.1007/978-3-319-20520-5_15

business. Design Futures acts as a collective placeholder for interdisciplinary design at Brighton and supports adaptive and open models for the curricula, which, through the development of teamwork and collaboration skills, prepares students to deal with a broad range of complexities in design research, design practice and design innovation.

Recent examples of our interconnected and outward-looking project teams include international experts, exchange students and academics, working alongside business partners ranging in scale from local start-ups to multinational companies. These project-based initiatives provide opportunities for internships and for longer-term strategies for collaborative partnerships.

15.2 A Short History

Design at Brighton The College of Arts and Humanities at Brighton has a long and distinguished history of design and of making that dates back to 1859. Design within the current college shares a creative environment with the broad range of humanities, media, film, music and the visual and performing arts (Lyon, 2009). In the context of the university and the critiques outlined above, the design team at Brighton have been experimenting with a range of different approaches under the collective banner of 'Design Futures'. These experiments are aimed at creating new, responsive, educational models informed by the histories and theories of design (Buchanan, 2001; Cross, 1999; Findeli, 2001; Frascara, 2007; Norman, 2010), that propose a reshaping of content, mode of delivery, and means by which these courses educate designers for the future. As a result, our design graduates are able to contribute valuable and meaningful insights, practices and skills to contemporary society (Fig. 15.1).

The emphasis is on finding pioneering ways to examine the relationships between images, spaces, objects and humans in order to interrogate and understand what knowledge they share. The model also seeks to explore how, through research and processes of synthesis, these insights can be applied to facilitate the generation of compelling ideas and meaningful innovation activities. Modules within each programme of study aim to develop a shared understanding of principles and methods that are teachable, comprehensible and transferable, as 'creative activities that generate intellectual value and are unique to the discipline of design' (Kolko, 2010). Our teaching is directly linked to our research and complemented by engagement with external partners. Modules and live project briefs support a distinctive and personalized student experience.

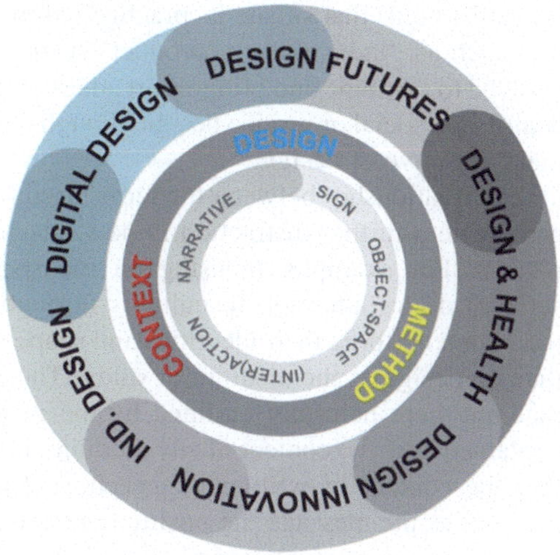

Fig. 15.1 To accommodate new forms of design specializations and hybrid/collaborative practices, the model of four orders of design thinking (Buchanan), less concerned with traditional disciplines and the way individual designers design, is adopted and updated as an outward-projecting model for interdisciplinary design

15.3 The Educational Mission

The concept of disciplines is a relatively recent phenomenon. The extent to which contemporary higher education has been quite literally constructed around the narrow but rigorous deepening of knowledge, reinforced through disciplinary specialisms, is surprising. Developments within disciplinary specialisms have enabled scientific research to progress rapidly and to advance human understanding across many fields. However, as a counterpoint to this, subjects such as design, which require the aggregation of different forms of knowledge, have had to regularly redefine their disciplinary identities as a consequence of the differential pace of change that occurs as different subjects incrementally advance and subsequently transform our understanding of the world and our place within it. This dynamic, although familiar and stimulating for practicing designers and researchers, posits particular challenges for design education. Challenges that until relatively recently have remained under explored.

There are many complex reasons for the lack of progress in design education in response to these challenges. Some obstacles are scholarly and others about the cultures of subjects. The design professions have evolved through the development of skills that have supported self-expression, self-criticism and self-reflection through means that are generally beyond text based practices. Sources of accessible, critically advanced and well-constructed scholarship have remained largely uncatalogued and invisible, limited by multiple digital technological challenges, economics and the capacity to systematically archive them for posterity such that richness of sensory knowledge is appropriately captured.

Within many professional contexts, design has become the common parlance for purposeful planning of systems, artifacts and services. Within the disciplinary structures of education, the term 'design' is generally prefixed by a range of specialist fields that situate its practice within contexts of specific forms of production such as, 'industrial' or 'product' design. More recently, these qualifiers have evolved to combine and bridge a broader range of thematic fields, such as design for health, and social or service design, though these too detract us from focusing on the meta-narrative of design.

The relationship between the term 'design' and its qualifying 'specialist' prefix can create a further distraction in how design practice is understood and communicated. For example, focusing on the specialist prefix (e.g. fashion, graphic, service) draws attention to cultures through and from which practitioners draw their content, forge their identities and shape their 'communities of practice', rather than on design as the medium of study. The interconnections between content and medium were helpfully outlined by the media theorist Marshall McLuhan, who explained how the content of any medium, is always another medium; for example, that the content of writing is speech, and of print is writing. Referring to 'the medium as the message' he positioned content (or subject prefix) by likening it to the 'juicy piece of meat carried by the burglar to distract the watchdog of the mind', diverting our focus from design as practiced or indeed from design education (McLuhan, 1964).

15.4 The Community

In the College of Arts and Humanities at the University of Brighton, design, design history, art and media programmes have been brought together to share a dynamic, creative environment along with architecture, interior architecture, fashion, textile design, graphic design, digital music, media, performing arts and photography. This has helped to create a highly charged creative environment with traditionally shared elements of study between different disciplines. However, these are increasingly questioned and challenged through new forms of collaboration between disciplines – not only across different cultural domains of creative disciplines, but also programmatically opened towards science(s) – environmental, social, medical, etc.

The college values and enjoys a diverse group of academics, practitioners, and students from a broad range of different backgrounds. In 2015 the College of Arts and Humanities supported approximately 4,200 students across Bachelor-, Master- and Doctoral- level studies, and employed approximately 270 academic staff. Academic staff within the university are supported to pursue research activities through Centres for Research and Development (CRDs), situated within each college. Within each CRD is a range of 'Research Initiatives' – specific areas of research interest, strength and opportunity that help to guide and coordinate research activities.

The interdisciplinary design research community is represented through the Design Research Initiative (DR-i). The mission of DR-i is to redefine and promote recognition of the fundamental significance of design research and to support research staff in their individual aims while encouraging collaboration with design researchers across many disciplines. DR-i is particularly concerned with the facilitation of innovative thinking and outlook across the design spectrum, initiating appropriate design research intervention in a wide range of fields, including national policy-making. It is therefore also deeply committed to the stimulation of sophisticated interpretations and understanding of design policy, practice and consumption. Its role is to advance the outcome of design research through a variety of mechanisms including knowledge transfers, external partnerships, publications, symposia, exhibitions and other forms of dissemination.

DR-i is integral to the operation, development and delivery of the interdisciplinary design strategy of the College of Arts and Humanities. Working closely with academic programme teams, it:

- **Acts** as an activator, and catalyst to trigger, develop and support research;
- **Creates** connections with regional/national/international funding bodies and industries;
- **Promotes** and develops research pathways and careers for both academics and students towards research project leadership, PhDs, post-docs and research fellowships;
- **Develops** its capacity and remit in acting as a bridge to external partners, projects, incubator spaces for design innovation and businesses in the UK and the EU.

As a research group, DR-i attempts to redefine and promote recognition of the fundamental significance of design, aiming to encourage collaboration with researchers across different disciplines. It is particularly concerned with the facilitation of innovative thinking across the spectrum, initiating design research in a wide range of contexts including the arts and STEM subjects (science, technology, engineering and mathematics). DR-i helps to facilitate the practical development and streamlining of UG and PG student research by introducing projects and themes with the objective to develop design research outcomes and collaborations at the highest level. As an example, the EU project FLAX has instigated material developments for interior and exterior applications: Design for the Body, Design for Well-being, Design History and Narratives. The project, which established a model for future research clusters, has created design research 'real-world' applications and developed new funding streams.

Although the term 'design thinking' is frequently used inconsistently among designers, theorists and writers, it can broadly be described as 'a creative mindset concerned with: how we conceive of, and understand, the world around us; how we consider and define challenges and solutions; and how we apply outcomes appropriately and sympathetically to requirements' (Ainsworth, 2013). Design thinking, as a connecting element of the community, has also begun to enable new forms of interconnectedness to be developed between community groups, processes, practices and theories of design with those of business and innovation. 'Design' therefore becomes the key element that

> "links creativity and innovation, such that it shapes ideas to become practical and attractive propositions for users or customers. Design may be described as creativity deployed to a specific end"' (Cox, George & Dayan, 2005).

Design research alongside the articulation of design thinking, and enacted through the practices of design synthesis, has ensured that the medium and methods have begun a progressive educational shift from invisible, discreet, internalized and personal processes to a more dialogic, collaborative and visible articulation of the act of design. This development facilitates greater engagement with other fields, communities and partners in the industry and education through which, and by which, design can be both understood and evaluated. It is unlikely that there will be 'a single "design thinking" toolkit . . . but in general, we need to know much more about the skills and knowhow that designers apply to their challenges and which ones can and cannot be transposed into different domains' (Hobday, 2012).

Our recent interdisciplinary research under the title 'Brighton Fuse' (http://www.brightonfuse.com) demonstrated the potential impact on innovation and economic growth from the fusion of creative, digital and IT skills within small and medium size firms. Included in its recommendations is also the development of new combinations of skills, knowledge and competencies that ensure that future graduates understand super-fusion and are able to recognize and sustain creative economies wherever they emerge. The Brighton Fuse also provided compelling evidence of the need for more responsive and socialized models of university education that not only transgress subject boundaries within institutions but also transform the delivery models and nature of knowledge exchange between business and higher education, and that value the integration of sensory and cognitive learning.

Our design and innovation community benefits from a location in a city with a highly successful digital media industry that includes world-class web, games and digital marketing companies. The fact that there are over 1,000 digital and technology companies, the biggest cluster per head in the country, has now been recognized by the government, which has strategically supported Brighton to become one of the three digital innovation centres in the country. According to recent surveys, around 23.600 people work in Brighton's creative sector, almost one-fifth of the local economy, and it has been growing at a rate of 22 % a year.

15.5 Teaching, Learning and Assessment

Debates on methodological practices in design are part of the familiar landscape of design research. The design of design education, however, has been less regularly interrogated.

Indeed, design in many industry or business contexts (generally presented under the umbrella of innovation or enterprise) has probably undergone more scrutiny than design in education.

It could also be argued, as Adamson (2007) has said of craft and its tradition, that design too has been beguiled by its own facility to seduce and to produce elegance at the cost of its scholarly development. In many educational curricula the scholarship and research of design have been eschewed, and this position defended as a means to preserve creative values of 'exploration', 'creativity' and 'play', with a fear that such practices may be stymied by scholarly engagement. As a result, undergraduate design education has developed a tendency to produce and sustain a series of self-fulfilling project-based conventions that restrict broader and more strategic critical reflection. In many cases, this has resulted in the separation of design practice from its research base and, arguably, from a more purposeful advancement of research in, through, and about design (Frayling, 1993).

Contemporary designers are employed to create interactions, services and experiences. They are also increasingly employed to design organizational structures to tackle complex social and political problems. However, as Norman (2010) suggests, designers 'have become applied behavioral scientists, but they are woefully under-educated for the task'. Instead, there appears to remain little within contemporary design curricula to provide the underpinning intellectual scaffolding or the tools, skills and knowledge to tackle our more complex 'wicked problems' with any degree of rigor or confidence.

While Norman succinctly identifies the broad deficits and challenges of design education, he offers little in terms of specific analysis or solutions. How we begin to resolve such dilemmas in the context of our rapidly changing world, and how we identify and revisit the essential skills and competencies required to undertake research, remain unanswered; as does the question of how this is achieved, in any chosen design sphere, without losing the richness, vitality and dialogue that shapes quality design education.

The dynamics of contemporary culture, the disaggregation of our individual identities, and our day-to-day encounters across physical and digital spheres present a wealth of new opportunities for the ways in which subject knowledge is gained and evolves. There is increasing value in the development of subject knowledge (e.g. graphics or interiors) as both content and medium transcend traditional subject thresholds and, instead, knowingly adopt models of 'communities of practice', in which the underlying principles and values of design and design education are more systematically and consciously shared and built upon. This could be achieved through rethinking its practices and their articulation, and by systematically embedding other ethical, moral and political forms of scholarship as core elements of design education, leading to a better understanding and alignment of design research, design practice and design education.

Like material to a craftsman, such opportunities could provide the means for students to iterate and to understand their relationship to the medium of design, prior to positioning themselves in more complex interdisciplinary environments. However, in an educational context, this requires both academics and students to overcome a second and vital transitional threshold: the shift from holding subject, specific specialist knowledge to becoming more intellectually and creatively agile, enabling the students to apply their skills in other fields, as well as through new forms of interaction and human association.

Developing an agility and understanding in students as to how and where to focus, transitioning between what is designated as medium and what is critical as content, and understanding that the content of any medium is always another medium, can help students in working both within the narrow confines of disciplines, whether theirs or those of others, and equally in working with a multitude of other specialisms. This is enabled through the process of shifting focus – distinguishing the 'enablers' (agents for any transformation) from the 'ennoblers' (values that mediate any given context).

15.5.1 Pedagogy

Underpinning our pedagogic developments in design education remains a firm understanding that design evolves from the intersection and fusion of different forms of knowledge. Such conjunctions offer opportunities to create hitherto unseen combinations of materials, techniques and technologies supporting product innovation, regeneration of urban or rural environments, and sustainable development.

The incremental transformation of the existing cluster of design disciplines, including fashion and textile design, and graphic and 3D design, started by creating Design Futures as a central hub of innovation and repository of interdisciplinary design thinking. Collaborative cross-discipline projects underpinned by research and links with the industry are intended to develop students as design pioneers and researchers, able to envision and locate new contexts for design as a profession and for the ways in which is design is conceived and practiced.

The underlying aim of our evolving curricula is to open up learning spaces capable of transcending traditional subject boundaries and to explore the core elements of the design process within a single framework. As a starting point we have provisionally adopted Buchanan's four orders of design: 'Signs, Things, Actions and Thoughts'. Student learning, which is built on the sequence of design experiences through the four orders, provides a basic understanding of how common critical content (meta-narrative of design) relates back to different media and professional contexts without a danger of compromising links to subject-specific disciplinary expertise. At the same time, this model of the curriculum design has the capability to help students to overcome the transitional threshold of the specialist knowledge to move progressively towards higher levels of interaction, narrative and content communication. The four orders of design are then folded into what we have called the GRID (Generative Research Interface Device), which is part of the development of a new conceptual toolbox to assist us with transforming our courses.

The concept of the GRID as an educational-developmental tool evolved as part of the research underpinning the development of modules which embody the pedagogic strategy of intersecting cross-disciplinary innovation (Kermik, 2013). Much like a map, the concept of the GRID was introduced as a metaphor for a network of concepts and methods, and it is employed as an abstract device for finding places and establishing connections in networks. Central to the idea of working with the GRID is to challenge the learner to develop a system of logic, and to define the rules and the script of the process of design. As a tool, it provides a structure to construct the language capable of communicating both the abstract as well as the intended specifics (route of discovery and exploration leading to proposition) – both intuitive and rational. In the context of interdisciplinary communication, the GRID serves as a lens, capturing the landscape or territory under scrutiny as seen from the perspective of different disciplines. Adaptable as a purpose-made framework, the GRID not only provides a level of accuracy in isolating a unique set of issues, variables and design opportunities, it also becomes a tool for creating feedback loops and making the design thinking visible.

The qualitative dimension of the development process and the ability to see design opportunities within and as part of larger systems emerge from this design space interwoven into the GRID. The methodology of finding and mapping connections and relationships as part of the system becomes a measure of depth and quality of the design thinking by picking up threads of narrative from that space, which is taken as infinitely dense. Balmond (2002) describes design space as a ‘black hole’, which contains all the possibilities: ‘to realize a design one has to map or unravel a trace from out of that hole. Each thread has its own story – we could leave it or stretch it out or fold it into our creativity.’ The GRID provides n-dimensional compatibility to record these unraveling processes as a narrative to be shared by participants within modules and projects and beyond disciplinary groupings.

Pedagogically, the critical importance of this school-and-college-wide design framework is to provide a questioning environment that will build on the distinctive and integrative strengths of design education to develop new formations of knowledge and ways of learning and researching. Commitment to learning is communicated as a shared collaborative process that includes both students and staff. The framework aims to equip students with the creative and intellectual skills needed to enable them to explore their chosen creative pathways with confidence and autonomy. It also seeks to assist students to understand relationships between fields of knowledge and to feel empowered to work beyond traditional disciplinary boundaries.

The following section provides three case study examples of projects and modules of study developed in the school under the common model of the Design Futures: ‘Socially Useful Design’, ‘The Human Body Form’ and ‘Material and its Form’. Each of these projects will be summarized and discussed in the context of interdisciplinarity at Brighton.

15.5.2 Socially Useful Design

Aims and Objectives: bring real-world relevance and methods of collaboration to academic studies;

Key Learning Outcome: understanding of the nature of systemic societal transformation processes;

Project Outputs: visual model of the systemic issues analyzed (2 x 4 meters), accompanied by a five-minute video outlining their solution.

'Socially Useful Design: A Systemic Approach' (SUD) is an example of the new pedagogic initiatives associated with Design Futures. Designed as a short project culminating in an interactive event, it brings together students from different courses from the College of Arts and Humanities and across the university to work together in an intense problem-solving environment and to share an experience of the application of design as a systemic interdisciplinary practice. The key goals of the project are to bring real-world relevance and methods of collaboration to the students' academic studies, and to develop in the participants an understanding of the nature of systemic societal transformation processes, the importance of systems mapping, and the necessity of identifying leverage points and design opportunities in such a methodology.

The project, which starts with a one-day primer session introducing the nature of a systemic approach to design and social change, involves students working in teams (and with external partners and stakeholders) in a three-day workshop event in an off-site location to enhance the conventional curricular routine of the students from a range of different programmes. The students, following a structured sequence of hourly and daily tasks and deadlines, work around a devised problem to create a large-scale visual model of a systemic social problem to which design solutions may be applied (Fig. 15.2).

The methodology of the workshop is derived in part from the Symbiosis in De-sign (SiD) systems innovation framework, developed by sustainable design agency 'Except Integrated Sustainability', and combines methodologies of collaborative innovation strategies such as hot-room teamwork, forms of complexity thinking, and systems analysis and implementation strategies for social problems. As part of the project, and as part of the learning experience for the 'hosting' group of students, the workshop space is transformed into its own universe, one that the students and staff inhabit as they pursue the project. This is achieved using work props such as furniture and lights. Project participants prepare and eat lunch together each day to enhance the bonding effect of the experience.

The project output includes visual models of the systemic issues analyzed (approx. 2 meters × 4 meters) and a five-minute video outlining their solution to accompany the models.

The social element of the project will greatly enhance the student and staff experience for all participants and act as an example of an exciting and unusual learning activity that happens as an extension of regular campus-based events. The

Fig. 15.2 Design at Brighton

event, which gives participants valuable experience of working in interdisciplinary teams, is an example of research-informed curriculum delivery, since the relationship between the programme of study and the partner institution is based on a shared exploration of design as systemic practice. Students will also be learning advanced research skills that they can apply to their own learning and practice. The workshop is a good example of an interactive experience concerned with the active co-production of knowledge and the fostering of a spirit of enquiry.

15.5.3 Human Body Form

Aims and Objectives: Facilitate personal reflection of 'disciplinarity' & collaboration

Key Learning Outcome: Development of interpersonal and interprofessional working skills

Project Outputs: Exhibition: Bodies Beautiful, Creative Campus Initiative, Jubilee Library, Brighton UK (2011).

Publication: Lyon, Ainsworth, Letschka, Haq., "An exploratory study of the potential learning benefits for medical students in collaborative drawing: creativity, reflection and 'critical looking'" BMC Medical Education Journal (2013).

The Human Body Form (HBF) is a drawing-based interdisciplinary project. Launched in 2009 and having been through a number of iterations, it brings together Medicine and Design students to explore the human form through drawing. Within the project, design and medical students are invited to participate in eight guided drawing classes to explore the human from through visual, sensorial, anatomical

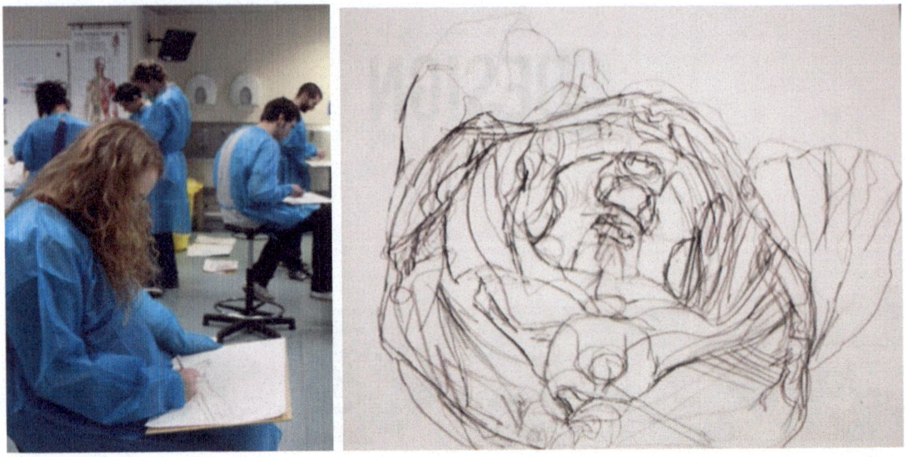

Fig. 15.3 Design students and medicine students drawing in the anatomy lab at the University of Brighton as part of the Human Body Form project

and perceptual domains (Lyon, Letschka, Ainsworth & Haq, 2013). Drawing, collaboration and facilitated discussion are key components of the project. Reflection and critical discourse are used at the end of each session when students are encouraged to consider what impact the experiences may have had on them, articulate their observations and insights with the group, and think about what effect these shared experiences may have on their future practices as either medics or designers. This introspective analysis encourages students to reflect upon their own disciplinary identity and its relationship to other forms of knowledge and specialist practice.

One of the aims of the programme is to take students to unfamiliar intellectual and physical places. In doing so, the edges of the disciplinary specialism within which they are being trained are brought into focus. Students are then encouraged to explore the similarities, differences and limitations of their own specialist knowledge in relation to those of other fields, and to consider what effects this may have on their own 'professional identity'. Within the context of interdisciplinary education, the aim is to enable participants to make informed decisions about the value of both disciplinary specialism and the shared knowledge of interdisciplinarity (Fig. 15.3).

15.5.4 Material and Its Form

Aims and Objectives: enable students to close the gap between the abstract conceptual speculation and reality;

Key Learning Outcome: ability to trace imagination and decision making in a feedback loop, helping the learner to progressively move closer to a clearer formulation of the design brief, its aims, a bespoke methodology and the intended outcome.

Project Outputs: process visualization based on the GRID methodology and a design proposition.

'Material and its Form' is an example of the experimental interdisciplinary modules with a degree of adaptability to accommodate a different annual focus or a theme. It utilizes the GRID as a generative device to enable students from different creative disciplines to develop and exchange ideas in the contexts of form, space and materiality. Typically, following the model of project-based conventions, designs evolve around ideas of abstraction and synthesis. Drawings and models – whether analogue or digital – employed in the creative process describe by default something else, another reality, whether intended as developmental stage recordings or annotations of geometry, scale or material specification for the design proposition.

Material and its Form has evolved into an annual project where various configurations of the GRID method can be systematically tested. Examples of the variety of investigations undertaken by students include narrative as folded 3D spatial construction; scripted landscapes based on text-based instructions and mathematical code; and concrete as 'liquid stone' – modular concrete building blocks derived from the set of variables including gravity and flow. Individual projects developed in Material and its Form extend the possibilities of applying research findings reciprocally to core studies, whether in architecture, design, fine art, photography or the humanities.

The primary purpose of employing the GRID tool in Material and its Form is to enable students to close the gap between the abstract conceptual speculation and reality. It aims to provide space for both the framework and substance of the investigation, and to ensure that design decisions are evidence-based. Rather than following a linear path towards a predetermined output, with the creative mind locked to a target, students are challenged to script a process as a way of finding a path from A to B with a level of reasoning where disciplinary stereotypes and personal ambitions of self-fulfilling seduction become irrelevant. The dynamic nature of GRID-based development traces imagination and decision-making in a feedback loop, helping the learner to progressively move closer to a clearer formulation of the design brief, its aims, a bespoke methodology and the intended outcome (Fig. 15.4).

Fig. 15.4 Material and its Form 2014 explored opportunities of working with materials considered as waste. Common nodes of connection to the GRID, which informed the outcomes of the project, included: reuse, regional, added value, production with limited resources and energy

15.6 Feedback, Conclusions and Future Evolution

Projects such as 'Socially Useful Design', 'Human Body Form' and 'Material and its Form' provide examples of how generative research methodologies can be applied to facilitate evolutionary models of design education. Feedback from participants identified a range of skills gained, such as the development of systemic thinking and strategies for integrated design practices. The relevance and coherence of the interdisciplinary engagement and the integration of projects with industry and community partners enable students to benefit from an extended range of disciplinary and professional practices. The application of research as an underlying imperative for curricula development and collaboration in the College of Arts and Humanities enables both students and academics to access different domains of information, pursue areas of interest in greater depth and contribute to knowledge fields not traditionally associated with design practices.

The relevance of disciplines and the reemergence of interdisciplinarity within contemporary university education brings a wealth of challenges and opportunities to design education in particular. The increasing ubiquity of information and the nature of design as a profession that brings together different disciplines and spheres of knowledge mean that design education is a crucial environment for supporting, facilitating and driving ongoing development. The role of the university in contemporary society is also drawn into question. Twenty-first-century life places new demands upon university graduates that go beyond the deep and rigorous knowledge gained in more traditional highly specialized subject disciplines.

Whilst deep knowledge and highly specialized learning is valuable, subjects such as design that thrive in the complex and vibrant areas where disciplinary specialisms meet, and which continually evade or distort convention and classification, present significant challenges for education. These challenges remain relatively underexplored. The design professions generally operate in practices beyond or, at least 'outside', text-based fields of knowledge. Critically advanced and well-constructed scholarship is therefore largely catalogued and invisible.

The qualifiers and prefixes often used to contextualize design practice, used in both industry and education, can also cause further distraction and difficulty in establishing and developing a 'meta-narrative of design'. Whilst scholarly debates on methodological practices in design are becoming part of the familiar landscape of design research, the same cannot be said of design education. The creative values of 'exploration', 'creativity' and 'play' are vigorously defended in many design curricula. This is perhaps under the seductive influence of design's own ability to produce beauty and elegance, and due to a fear that such freedoms may be compromised by scholarly engagement.

The demands of contemporary design professions, however, require a more complex and rigorous set of skills that enable graduates to engage with and resolve 'wicked problems' with competence and confidence. There is, therefore, a need to rethink the ways in which design competencies are practiced and articulated. The systematic embedding of other ethical, moral, social and political forms of scholarship could also form core elements to assist a realignment of design research, design practice and design education. This shift, however, would require both students and academics to willfully develop their skills alongside other fields and explore new forms of interaction and human association.

The College of Arts and Humanities at Brighton has developed a range of different experimental approaches to interdisciplinary design education under the overarching concept of Design Futures. The aim is to develop new ways of educating learners that will enable future generations to contribute valuable and meaningful insights, practices and skills to contemporary society. Each experimental programme of study is linked directly to our research practices and engages with external partners. The coordination of the framework enables a range of projects to be developed that provides a questioning environment for learners to explore and in which to develop new formations of knowledge and new ways of learning and researching. The aim is to inform and empower students to become autonomous as practitioners in their chosen field and to not feel constrained by traditional disciplinary boundaries.

The three case study examples outlined in this chapter provide a brief insight into how the ideas presented have been developed and applied through a different focus on the nature and practice of interdisciplinarity, and how each seeks to have a different kind of impact on student learning. The Socially Useful Design project adopts a co-production model and seeks to create a real-world design scenario in which student groups from a range of disciplinary areas come together to tackle research-informed societal problems. The Human Body Form project adopts an analytical and introspective model of interdisciplinary working that seeks to promote awareness of the value and limitations of disciplinary specialisms. Material and its Form encourages participants to use the idea of 'material' in its broadest sense as a means to explore relationships between abstract conceptualization and physical realities.

Whilst practices in design education will undoubtedly remain contested, it is the vibrancy of constructive debate that enriches and enhances the quality of outcomes. For design education to evolve in a way that enables it to equip students with the competencies required for success in future societies, more needs to be done by design scholars to rigorously test and challenge established norms in design education and to test and evaluate innovative and experimental approaches to teaching and learning.

Chapter 16
The EIT Digital Master School: A Program to Foster the Education of Entrepreneurial, Innovative and Creative Students

Maurizio Marchese, Lena Adamson, Carl-Gustaf Jansson, and Anders Flodström

16.1 Preamble and Short History of EIT Digital

The EU has 500 million citizens, and is the largest economy in the world. Still, Europe is not a global leader in innovation. In China you always meet students with a sense of urgency to finish their education to get out into world and to make a difference and to create value for themselves and for others. In Europe you seldom meet the same spirit. Worse, the educational programs do not help the students at European universities to gain that spirit. The universities in Europe are focused on research as their main task, forgetting that the European university's main "raison d'etre" is to foster the new generations of young Europeans that will make Europe the most innovative part of the world and a Europe filled with entrepreneurs that renew science, business and society.

In 2008, the European Parliament founded the EIT (European Institute of Innovation and Technology), to answer the need to increase Europe's competitiveness on the global market. EIT is a distributed organization consisting of a Governing Board, Headquarters in Budapest, Hungary, and at present three Knowledge and Innovation Communities (KICs). These are multi-stakeholder, independent, legal and financial integrated entities, governed by a CEO appointed by a

M. Marchese (✉)
Department of Information Engineering and Computer Science, University of Trento, Via Sommarive, 9, 38123 Povo, Italy
e-mail: maurizio.marchese@unitn.it

L. Adamson
Stockholm University, Stockholm, Sweden

C.-G. Jansson
KTH Royal Institute of Technology, Stockholm, Sweden

A. Flodström
EIT Digital, Brussels, Belgium

B. Banerjee, S. Ceri (eds.), *Creating Innovation Leaders*,
Understanding Innovation, DOI 10.1007/978-3-319-20520-5_16

board of main stakeholders from academia and business. Three areas – climate, energy and ICT– were first identified, and in January 2010, three theme based KICs were in place: ClimateKIC, KIC InnoEnergy and EIT ICT Labs, the latter with the mission to drive European leadership in ICT innovation for economic growth and quality of life. Currently EIT ICT Labs has recently changed his denomination into EIT Digital and includes seven Partner/Nodes, in Berlin, Eindhoven, Helsinki, Paris, Stockholm, London and Trento and two associate ones in Budapest and Madrid. EIT Digital has created in these nodes Co-Location Centers where local innovation ecosystems are nurtured and research, education, innovation and business are highly integrated, in strategy, in execution and between stakeholders.

Innovation drives the global economy and innovation changes the ICT professional roles very fast. In fact, so fast that the higher education in Europe must change and encompass a curriculum where generic skills are given more emphasis to handle the dynamism. ICT innovations also both make jobs obsolete and create new ones and higher education plays a decisive role to make new graduates and (young) jobseekers employable. A higher education revolution is needed where higher order cognitive (soft) skills, multidisciplinary thinking and learning by doing guide the design of educational programmes.

A recent US survey asked, if the US bachelor and master graduates in engineering are ready for the jobs they been educated for; 72 % of the faculty, 45 % of the students and 42 % of the employers answer yes. Clearly a majority of the "consumers" of the education answer no to the posed question. An analysis reveals that students and employers consider that graduates do not have the creativity, innovative and entrepreneurial skills to work in a company context and use the knowledge they learned. The results corroborate the global trend that we move from a knowledge society to a competence society where the understanding of knowledge and the ability to apply knowledge is a necessity. For this transformation of higher education to happen a joint stakeholder perspective from faculty, students and labour market must exist and must be applied and all stakeholders must have operational roles. Faculty and students do also disagree upon the reasons behind low retention (high drop out) rates. 32 % of the faculty thinks it is due to that subjects are too difficult for the dropouts while only 9 % of the students agree. Students believe that lack of motivation is the key factor.

Do we educate the right students in the right way to create the technically skilled, innovative and entrepreneurial workforce with a sense and spirit for new business that is needed to transform Europe's way of doing things in industry and in society? We work within a global and European infrastructure of Universities and the students' success and legitimacy should be secured by national quality assurance and degree systems. However, these are created and adapted mainly to the faculty's view of a relevant and high quality education within a certain topic or area.

EIT was created to make new education happen: faster and with higher quality in Europe than anywhere else. This demands a certain urgency and at a systemic approach that is sustainable and can be scaled up. The EIT Digital way of educate should be a Yes to the question posed above and the way we validate the Yes is through applying the EIT Label quality system and adhere to the learning outcomes that form the objectives of our educational programs.

16.2 Program at a Glance

The job market asks for academic professionals, with higher order creativity, value judging and leadership skills that will produce innovative and entrepreneurial engineers and researchers. It is clear that "start-ups" created by academic entrepreneurs will play an important role in generating new jobs in the coming years, while large enterprises will focus on core business and will certainly need employees with new competencies but will not increase, rather reduce their number of employees.

The educational challenge is to educate enough professionals in the ICT field with the soft skills needed. EIT Digital initiatives in ICT education reflect an innovative approach to provide flexible, blended learning paths for the new "skilled entrepreneur" professionals. An example of such an initiative is the EIT Digital Master School: a novel portfolio of Master level curricula (defined through an agreement in 2011) offered by a consortium of EIT Digital and 21 leading European universities. It is based on the concept of a 2-year Master with a technical major of 90 ECTS and an Innovation and Entrepreneurship (I&E) minor of 30 ECTS. The specific programs cover topical and emerging ICT fields such as Service Design and Engineering, Security and Privacy, Internet Technologies and Architectures, Human Computer Interaction and Design, Embedded Systems, Digital Media Technologies, Cloud Computing and Services and finally Data Science[1]. The schematic layout of the educational programs is described in Figure 16.1.

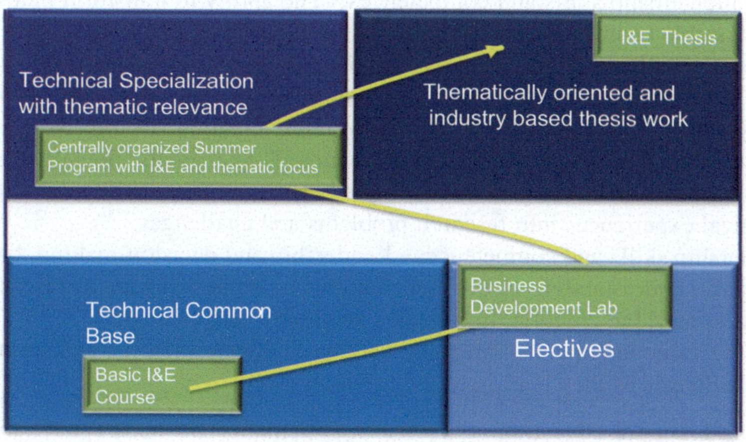

Fig. 16.1 Schematic layout of the educational program EIT Digital Master School

[1] The details on the technical majors and their full curricula can be found at http://www.masterschool.eitictlabs.eu

Students taking part in these programs can thus acquire in-depth, hands-on skills at the cutting edge of scientific and technological domains, studying and working in a mixed academic, industrial and start-up environment. Having welcomed the first cohort in 2012, the master school is currently recruiting its fourth students' cohort.

16.3 Educational Mission

EIT Digital embrace an educational vision, to make European work force more effective, innovative, entrepreneurial and competitive. The new higher education graduates should be able to understand and grip the industrial and societal challenges we foresee and to create a better life for the Europeans. In an increasingly dynamic scientific, technical and human development they should understand that lifelong learning is a prerequisite for the success.

The basis of EIT Digital educational vision is the knowledge triangle i.e., the seamless integration of research, education and innovation and to integrate the perspectives from public and private stakeholders in Europe's future, to form the European competence society needed. Towards this end a set of overarching learning outcomes were initially formulated for all EIT KIC master programmes, these master students should be able to demonstrate:

- Creativity skills and competences - The ability to think beyond boundaries and systematically explore and generate new ideas.
- Innovation skills and competences - The ability to use knowledge, ideas or technologies to create new or significantly improved products, services, processes or policies or new business models.
- Entrepreneurship skills and competences - The ability to transform innovations into feasible business solutions.
- Research skills and competences - Knowledge and understanding of cutting-edge research methods, processes and techniques; their application, within their study field; the investigation of new venture creation and growth, and the capability to work in cross-disciplinary teams in the thematic field of their KIC.
- Intellectual transforming skills and competences - The ability to transform practical experiences into research problems and challenges.
- Leadership skills and competences - Leadership and decision-making, based on a holistic understanding of the contributions of higher education, research and business to value creation, in limited sized teams and contexts.
- Making value judgments - An appreciation of ethical, scientific and sustainability challenges as they pertain to their field of work.

The EIT Digital education adheres to a systematic "School&Tools" approach providing blended (mixed Cyber and Campus environments) programmes that deliver T-shaped talents who are able to combine deep technical ICT knowledge with broad entrepreneurial skills. The "T-shaped" metaphor refers to professionals with deep skills and expertise in a single technical field as well as a set of broadly applicable non-technical abilities, e.g. related to innovation and entrepreneurship or to collaboration and communication. The Master School- but also the EIT Digital Doctoral and Professional Schools - build their programmes on “Tools”, such as EIT Digital partner university education programmes, Co-Location Centres and Online Learning Platforms.

The challenge is to be able to train an adequate number of professionals in the ICT field. In order to succeed in this, there is definitely a need for structural initiatives at a national and European level based on both a correct, more engaging school orientation and on growing the educational offering from universities. Advanced education strategy within the EIT Digital has on the other hand focused on exploring innovation both in terms of methodology and contents for training programs at master and doctorate levels. The focus is therefore mainly on the innovative dimension and on courses quality improvement rather than on the number of students (for which the structural measures mentioned above are necessary).

To this end, current EIT Digital initiatives in ICT Education reflect an innovative approach aiming to provide flexible, blended learning paths for the new “skilled entrepreneur” professionals.. New ICT professional roles and jobs share many traits with those described above - for example Security and Privacy experts, Big Data analysts, Cyber-Physical System architects, Urban Systems architects, e-Health architects. They have deep technical ICT expertise along with generic skills and understand the specific infrastructure and application areas.

16.4 The Community

EIT Digital Master School is an exclusive joint initiative by the leading technical universities and business schools in Europe (21 universities in total) with additional mentoring and partnering from excellent European research organizations and leading business partners. Currently the following universities are contributing to the Master level programs: Aalto University, Turku University, Abo Academy, KTH, ELTE, Budapest University of Technology and Economics, TU Eindhoven, TU Delft, Twente University, Saarland University, TU Darmstadt, TU Berlin, University Pierre and Marie Curie, University Paris-Sud, Rennes 1 University, University of Nice Sophia Antipolis, Institute Mines-Telecom , Trento University, TU Milano, University College London and Madrid University. In every university, a coordinator is appointed for each programme of the Master School. In 2014/15 the MS enroll 400 students. The goal is to scale up to 1000 students in a steady state.

The Master School employs a centralized admission, where all applicants are evaluated according to a standardized evaluation model. Individual admissions and scholarship allocations are strictly based on an individual quality score. We have been able to recruit high quality and innovative students as shown by the low number of drop outs (high retention); in average 10% for the first three cohorts.

The students innovative and entrepreneurial qualities have been proven by the many innovation and potential start up ideas they have undertaken.

For the three first three cohorts the global number of applications has grown as follows: 500->800->1500. The students are mainly men (70 % in the 2014 cohort) and non-Europeans (65 % in the 2014 cohort). Two important recruitment challenges are to reach balanced (50/50) cohorts concerning female/male and European/non-European students.

The Alumni is important for branding the EIT Digital education. The Alumni carries the characteristics of the EIT Digital education to European companies and labour market. An EIT Digital Alumni foundation/association was created in 2014. The present Alumni foundation/association is made up of the first Master School graduates. The Alumni foundation can appoint earlier executives and faculty that have dedicatedly worked with EIT Digital as Alumni.

16.5 Methodology

The EIT Digital Master programmes are based on the common framework of the EIT Quality Assurance and Learning Enhancement (EIT-QALE) model. The model is based on the learning outcome paradigm as it has been brought forth within the Bologna process where the aim is to move from 'teacher driven' to 'student centred' teaching and learning; changing higher education from being just knowledge based into also being competence based. The quality assurance has a strong focus on the promoting and enhancing aspects of quality, in addition to that of accountability. The main question to answer is "do programs ensure that students achieve the EIT learning outcomes?" That is, that the programs provide students with opportunities to develop a true entrepreneurial mind-set combined with knowledge triangle skills and competencies.

The structure consists of a total set of five quality indicators each comprised of a number of assessment fields where the first, Indicator 0, consists of Obligators (such as mobility windows, number of ECTS, etc.) that must be fulfilled before any further evaluation work is proceeded with. For the accreditation/labelling process of new EIT programs the following indicators are used:

1. Quality Indicator 1 – Aligned teaching and content coverage and
2. Quality Indicator 2 – EIT learning environment and facilities are used.

Indicator 1 uses five different assessment fields to evaluate if the program sufficiently covers the EIT learning out-comes in relation to the thematic field of the KIC, if is characterized by aligned teaching and activating teaching methods and if it provides students' access to grading criteria (rubrics).

Indicator 2 is concerned with the study environment in terms of "robust entrepreneurship education", innovative "learning-by-doing" curricula, mobility and the European dimension and openness to the world. This part of the model has been used in all three EIT KICs by now to label all new programs (see Fig. 16.2 for more details).

Q Indicators: ---------------- Assessment areas:	*Q Indicator 0 COMPULSORY REQUIREMENTS*	*Q Indicator 1 ALIGNED TEACHING AND EIT CONTENT COVERAGE*	*Q Indicator 2 LEARNING ENVIRONMENT AND FACILITIES*	*Q Indicator 3 RESULTS*	*Q Indicator 4 STAKEHOLDER EXPERIENCES*
field 1	0.1 Mobility	1.1 EIT Overarching Learning Outcomes Coverage	2.1 Robust Entrepreneurship Education	3.1 Student Creativity	4.1 Students
field 2	0.2 Business Partner Curriculum Collaboration	1.2 General Quality of Intended Learning Outcomes	2.2 Highly Integrated, Innovative "Learning-By-Doing" Curricula	3.2 Achieved Learning Outcomes	4.2 Alumni
field 3	0.3 ECTS, DS and Recognition	1.3 Fit for Purpose Assessment	2.3 Mobility, European Dimension and Openness to the World	3.3 Retention Rates	4.3 Other Stakeholders
field 4	0.4 Application, Selection and Admission	1.4 Grading Criteria		3.4 R & D Projects on KIC Educational Activities	
field 5	0.5 English as teaching language , EIT Logo	1.5 Active and Appropriate Teaching Methods			

Fig. 16.2 The EIT Quality Assurance and Learning Enhancement Model

For the future review process of on going programs there are two more Quality Indicators focusing on the results of the programs. Quality Indicator 3 – Results consists of four assessment fields. The first field evaluates students' creative thinking and potential and the second their achieved learning outcomes. These will consist of samples of actual (degree) products by EIT students. The third assessment field of this indicator consists of retention rates. In the case of low retention this needs to be closely analyzed since student drop out does not automatically mean low program quality. The fourth and last assessment field concerns outcomes by the KICs in the form of published articles, reports, conference presentations etc. on research and development projects on KIC educational activities. This assessment field will stimulate the KICs in doing close evaluations and research on their educational activities in order to know what results they achieve and why. This assessment field will truly drive the quality of these educational programs in the sense that it will promote researchers to keep their "research glasses" on also in the teaching context. Hence it will enhance the teaching research nexus in a concrete manner and contribute with new knowledge in the field of teaching and learning in higher education.

The last indicator Quality Indicator 4 – Stakeholder experiences is divided into four assessment fields, stakeholder experiences and opinions of a) students b) alumni c) industry/business stakeholders and d) other stakeholders. Data will be gathered by questionnaires or interviews depending on how big the groups are. The questionnaires will primarily be on issues to do with Indicator 1 – 3 in order to keep the model focused.

The EIT Digital Master programs integrate a number of specific features into their syllabus that add to its overall innovative aspects. More specifically:

- A high-level, intensive program of courses on enterprise and technological innovation - the so-called **Innovation & Entrepreneurship (I&E) Minor** - which introduces elements such as business plan development, market analysis, teamwork in International and distributed environments to the students' competencies. The I&E Minor is organized in four main modules and are organized as follows: (1) Basic Course on idea generation, technology-based entrepreneurship, marketing and markets, organization and project management, new product and process development, entrepreneurial finance, human resource development; (2) Business Development Laboratory on business plan development in phases – (a) idea recognition – (b) concept design – (c) Business Plan conceptualization – (d) Business Plan presentation, application of subjects from basics supplemented with business ethics, IP Management, and market research; (3) Summer Program on I&E on specific technical themes proposed by EIT Digital Action Lines (i.e. research and business areas) where to work in teams on business ideas creation and pitching: (4) I&E Minor thesis where students apply, synthesize and evaluate prior learning in the context of a specific technology and business case.
- **Geographical Mobility:** up to now, geographical mobility for European students has mainly happened through exchange mobility programs backed by bilateral exchange agreements between individual universities and supported by EU through Erasmus initiatives. The number of "free movers" within Europe has increased, especially within areas of economy, financial and management studies i.e. the MBA sector, but not significantly within engineering. The EIT Digital Master School programs with the objective to recruit bachelors broadly over Europe have to overcome the localization barrier. To this end a specific component of the Master program has been added, namely the mandatory mobility between at least two different universities within the consortium in different countries.
- A **final thesis project** carried out in one of the many EIT Digital consortium's companies, such as Nokia, Siemens, Philips, Telecom Italia, SAP et al. To this end a specific component of the educational program has been added, namely a mandatory organizational mobility, meaning that an internship period for the students in a non-university organization is required, typically integrated with the thesis work.
- **Participation in events** organized within the European consortium such as summer schools, winter schools, international competitions and specialist workshops where the students can work together in interdisciplinary and international groups to apply and evaluate their competences through a number of learning by doing exercises.

16.6 EIT Digital Summer Schools

As an example of both an I&E and event educational components we will shortly describe the two Summer Schools organized by EIT Digital Master School, held in Trento in 2014. The aim of both Summer Schools has been to teach participants basic I&E and business skills, to enable them to perform both customer and business development process on their own innovative idea in respectively (1) a Cyber Physical System (CPS) and (2) in a Security and Privacy context applied to a societal relevant thematic area. During two weeks, the students gathered in teams generated and refined their own ideas, pivot when needed, and developed their own seminal ideas until becoming well contextualized business ideas. Any team developed during the two weeks a clear description of their innovative idea and the problem they were supposed to solve, the relevant market, competitors and unique sales proposition, business model, cash flow and needed competences to execute the plan. On the last day each team presented their own idea in front a business panel (see Fig. 16.3).

Both schools have been organized in two main sessions. Within the school, business challenges has been introduced on the following themes: (i) embedded systems, (networked) control systems and wireless sensor networks and (ii) - Privacy-Aware and security in users' digital life. Students have received interactive teaching and tutoring on both business and technical aspects. Each student group

Fig. 16.3 Life at EIT Digital Summer Schools

has created and developed their own innovative project proposal addressing one of the particular societal challenges they consider as relevant. Through teamwork, students have also performed market studies, analyzed competitors, defined business models and return of investment and explored social and usability aspects of the proposed application. Finally, each student group has jointly presented its project to an evaluation panel.

The main features embedded in the design of the program were:

- Stimulate students to generate ideas individually while exposing them to the first technical lectures. The first lectures were designed to give participants and overview on the market, the best practices and future trends.
- Motivate all students in presenting their own individual ideas to the rest of the class. From that moment onwards the "idea trading" phase started. Ideas considered by students as valuable attracted students whose ideas were not considered relevant.
- The idea-trading phase ended with the team-forming deadline. Team forming was subject to rules designed to maximize diversities within any team.
- From that moment onwards (day 3) students started developing their own ideas while being exposed to other I&E lectures and tutorials, S&P domain lectures and use cases.
- All teams were subject to two sessions of pitch clinics to strengthen their public speaking skills
- Furthermore all teams were subject to a project surgery session where mentors critically analyzed all documents leaving one more day to tune presentation and report before the finals.

Moreover, the educational activities were organized around the following key dimensions: *Working on Individual Ideas*; *Presentation of Individual Ideas*; *Idea Trading*; *Team Forming*; *Coaching & Check Phases; Pitch Training*; *Project Surgery*.

16.7 Lessons Learned and Future (R)evolution

The EIT Digital Master School is based on a clear mission and on a framework/infrastructure to execute that mission. EIT Digital is on its way to realize the mission but need to further improve the quality of the I&E education and the integration of the I&E minor with the technical major. A tool to do this will be to use blended format teaching and using online modules to harmonize and increase the quality of the I&E education between different programmes and locations.

16.7.1 The Lessons Learned

3 years of operation of the Master School should be seen in the perspective that we have been building up a novel European education, where new skills and competencies related to creativity, innovation, entrepreneurship and leadership have become integrated parts of the technical programmes. A new breed of engineers with the generic competences necessary to handle innovation and business cycles that are shorter than educational cycles has been fostered. EIT Digital has done this, not unblemished but still successfully. The main lessons learned so far are:

- The programme coordinators must be idea driven and motivated in the execution of the programmes. They must assume the complete responsibility for the full (technical and I&E) programme.
- The programmes must be taught by a faculty with knowledge of the basic ideas for EIT Digital education. EIT Digital must become better in communicating with its teaching faculty at the partner universities.
- There is a need of a recognized and agreed internal quality assurance system that guides the educational work.
- The industrial (labour market) perspective on the curricula and the learning outcomes must be bettered cared about.

16.7.2 The Sustainable Future

Sustainability has other dimensions for education as compared to innovation and business development. What makes mankind sustainable is education! When **Homo Sapiensis** defeated **Neanderthals** it was because they taught their offspring effectively what they learned and experienced themselves. i.e. mankind has social genes that enable us through education to change fast and in a constructive way.

EIT Digital is at present funded by EU under Horizon 2020. The EIT regulation from 2008 makes it clear that the operational arms of EIT, its KICs, should have a limited lifetime concerning EU funding. 10 to 15 years is mentioned. A build up, a peak and decline phase of the funding are anticipated. The EU funding should be seen as an investment in the KICs and the KIC should create business models within innovation and education that leads to financial sustainability.

Higher education and the European universities are close to 100 % taxpayers' financial responsibility as opposed to the US where tuition fees paid by students and endowments to universities from wealthy individuals, organizations and companies play important roles. EIT Digital task is **not** to create a parallel alternative to the European Universities when it comes to higher education. EIT Digital should be instrumental in developing higher education in Europe and bring in new added values to education at European universities.

A scaling up (dissemination) phase should have the objective that universities create and assume financial responsibility for programmes with the EIT Label.

Programmes integrating technical content with teaching generic (cognitive) skills will become the role model for a majority of the education at European universities. Educational development is open ended, we can already foresee changes in the way EIT Digital educate. Online blended and MOOC educational tools often enable the changes. Professional learning in Europe will depend on these changes to make it a true lifelong learning for most professionals.

A sustainable model for EIT Digital education would then be that new pedagogic and work models pioneered by EIT Digital leads to changes in the way European Universities educate and make Europe an attractive place for global talent. This is also in line with the EIT mission to change European Higher Education. EU, European nations, universities and companies should pick up the price tag for EIT Digital.

In Europe higher education and its importance for future generations is not high on the agenda for wealthy people or companies. Higher education is considered to be a taxpayer paid activity. Europe has a number of major companies (many of them being EIT Digital partners). Their future success depends critically on the access to employees with the right skills, technical and cognitive. There is a global lack of young and older people with the right competences. In Europe between 400 and 900 thousand within the ICT and ICT intense sectors for the next years as stated by OECD. EIT Digital plays a pioneering role in creating this new breed of engineers and industrial PhDs. European companies should invest in higher education for the sake of their own sustainability. Industrial and societal stakeholders in European higher education should donate to the EIT Digital "*Higher Education for future European Innovators*" foundation with the objective to give the best global STEM talents the scholarships to study at EIT Digital programmes in Europe.

Chapter 17
Making an Interdisciplinary Difference: Twenty Years of Design, Business and Technology at Aalto

Mikko Koria

17.1 Aalto IDBM at a Glance

The International Design Business Management (IDBM) program is an interdisciplinary offering of Aalto University in the Helsinki area, Finland. Since 1995, it has delivered world-class multidisciplinary and systemic research and learning in global business development through design and technology. The aim is to educate global producers and leaders of innovation in new product, service and business development.

The program builds on the premise that new wealth, meaningful social innovation and solutions are increasingly generated in the spaces between disciplines and thus there is a need to educate interdisciplinary professionals. Furthermore, the program understands that students originating from different disciplinary backgrounds have distinct worldviews, capabilities and skills that are linked to their institutional backgrounds; this variance underpins the creative abrasion that enables innovation. Lastly, the program builds on the wide idea that joining relevant design, technology and business learning with the creativity that exists in all individuals unleashes the full potential to create meaningful innovation in business and society (Fig. 17.1).

IDBM is not a design management program (although that is a part of the curriculum). It is also not only educating individual designers, technologists or business people, but also bringing them together that makes a difference. We characterize the program as being a platform to educate innovation producers and changemakers who can make a difference in this world through interdisciplinary approaches.

In practice, the IDBM masters degree program is a two-year, 120 ECTS (European Credit Transfer System) joint offering between the Aalto University

---

M. Koria (✉)
Department of Management Studies, Aalto University, P. O. Box 21230, 00076 Aalto, Finland
e-mail: mikko.koria@gmail.com

B. Banerjee, S. Ceri (eds.), *Creating Innovation Leaders*,
Understanding Innovation, DOI 10.1007/978-3-319-20520-5_17

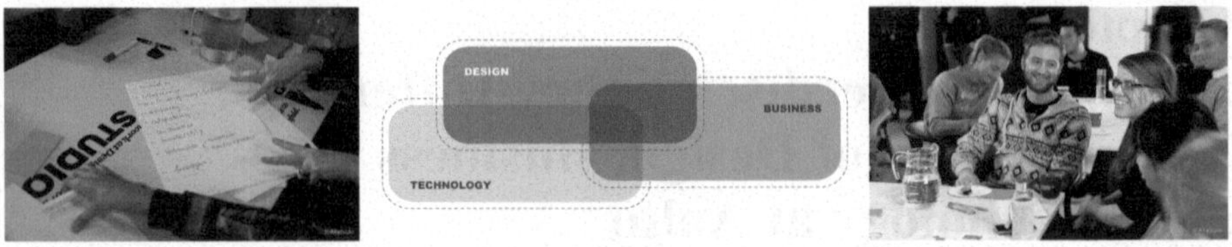

Fig. 17.1 At the junction of Design, Technology and Business since 1995. Akatsuki Ryu photos copyright used with permission

Schools of Art, Design and Architecture, Science and Business. The students come in equal numbers from design, technology or business backgrounds, with an annual intake of up to 40 full master's students, supplemented by minor studies and exchange students who bring the total up to a maximum of sixty participants. The minor studies are open to students in any relevant masters program.

While the IDBM master's students will graduate with degrees of Master of Arts (MA), Master of Science in Technology (M.Sc. (Tech.)) or Master of Science in Economics (M.Sc. (Econ.)), depending on their first degree, they all undertake the same core courses in design, technology and business during their first year.

The core offering of the program includes an eight-month long industry project with a real-life business enterprise setting, with small (four or five student) multidisciplinary teams that are balanced in terms of business, design and engineering students, coached by multidisciplinary faculty and expert industry tutors. The student teams develop new product, service and business concepts for the real world, using problem- and practice-based learning methods.

The central set of seven courses in design, business and technology support the industry project work in the first year. During the second year relevant elective course work is undertaken based on the personal interests of each student, which might include exchange programs and courses in other universities or complementary disciplines within Aalto University. The program has double degree tracks with Tongji University and ESADE business school, in addition to the extensive university-wide exchange programs. The second year also includes a 30 ECTS master's thesis.

Since 1995, IDBM has educated over 800 master's level students from 25 countries, with 190 completed industry projects with 120 partners of both Finnish and foreign origin. Within the program, attention has been directed towards the creation of an IDBM community of practice (through the IDBM Club structure), which has helped to link alumni to the program in a loose and informal fashion.

Over the years, Aalto University Executive Education has run design business modules in both professional development and for eMBA courses, based on the master's program.

17.2 Interdisciplinary Approaches to Higher Education Since 1995

Aalto University was created in 2010 by merging the Helsinki School of Economics, the University of Art and Design Helsinki and the Helsinki University of Technology. This merger represents the coming together of three top institutions of higher education, each with a long history of research, education and societal impact in the Finnish context. The current structure of the university is based on six schools: Business, Arts, Design and Architecture, Science, Engineering, Chemical Technology, and Electrical Engineering. The Aalto University community consists of twenty thousand students, five thousand staff members and close to four hundred professors.

During the merger process the legal structure of the university was built on the basis of a public-private foundation, with an endowment coming from both the industry and a matching multiple from the government. The transformation of the three hitherto public universities into a single public-private foundation has had deep implications in terms of the governance and administrative practices, not forgetting the distinct operational cultures that had been built up over the last century.

The International Design Business Management (IDBM) program was set up in 1994–95 as a response to the needs of industry. At the time, the business field was, largely speaking, not very aware of the full range of opportunities that design could create for them; at the same time the various areas of design felt that business was not really accommodating their specialist knowledge and nature in the best possible way, especially in terms of leveraging design on a more strategic level. Service design was not yet on the agenda, and the focus was on products – the concept of designing business models was similarly not yet on the radar.

Public funding from the Finnish Ministry of Industry and Trade was obtained to run the program for the first few years. The program started as a minimum viable concept, with a Program Director and Course Coordinator, and attracting the first paying business enterprises to join required extensive effort. The first few rounds quickly proved the value of the concept and over the first five years the program gained a reputation for excellence and for bringing the real-world into academia through close collaboration.

Initially IDBM was built up as a 24–30 ECTS minor studies program that enhanced master's courses in engineering, business and arts. The industry project, together with a design business management course, formed the original building blocks, and students could select further courses from the offering of the three universities. At the time, this represented a major breakthrough in student mobility.

The IDBM minor studies program was one of first (if not the first – the author has no knowledge of programs before 1995) multidisciplinary offering in the world that emerged from joining design studies with business and technology. This three-way arrangement between three universities was quite visionary at its time, and hard to implement initially. Still today almost all of the collaborative and interdisciplinary programs are created with a single institution and/or between two schools – balanced tripartite agreements are rare, especially degree-awarding ones, due to inherent complexity.

The delicate balance between the disciplines has been the cornerstone of the program since the beginning and remains one of its distinguishing key factors. Anecdotally, it can be observed that the program was built on "design thinking" principles much before the concept became widely diffused (of course one can think that "designerly" ways of thinking have been around for a very long time). The program also predates all of the current programs that are discussed in this book.

The IDBM minor studies program is still offered today, and has been somewhat expanded to include also a core course on creative teamwork, together with electives from the Aalto schools and other institutions of higher education.

17.2.1 From a Multidisciplinary Minor to an Interdisciplinary Master

A key development came to light in early 2008, when a decision was made to merge the three universities behind the IDBM into Aalto University. The same person (the then Rector of UIAH) was behind suggestions for both the IDBM in 1994 and the coming together of the three best-in-kind Finnish universities in 2008. While it would be totally presumptuous to argue that the IDBM program created the situation in which three universities would merge, the early excellence of the program was seen to be a role model for multidisciplinary academia-industry collaboration.

The IDBM minor program was requested to develop a master's-level program that would build up experience for a full degree initiative that was based on interdisciplinary and practice-based approaches. The work took two years, and despite the goodwill of all parties, it took significant effort to overcome the administrative challenges that the three operational cultures and ways of thinking imposed on the initiative.

The operational model was built on university-level funding and by reporting to the vice-president of the teaching activities. This bypassed departmental and school-level politics (to which many interdisciplinary programs fall victim) and allowed for an initial five-year development period. Concurrent with the development and launching of the IDBM degree program, Aalto University also embarked on building three other interdisciplinary platforms: The Design, Service, and Media Factories (Fig. 17.2).

As a later development, Aalto has seen student-driven entrepreneurship grow exponentially over the last five years, and one of the prime European start-up events, Slush, is organized every year in November, bringing together over ten thousand global participants. The university has been proactive in 'not getting in the

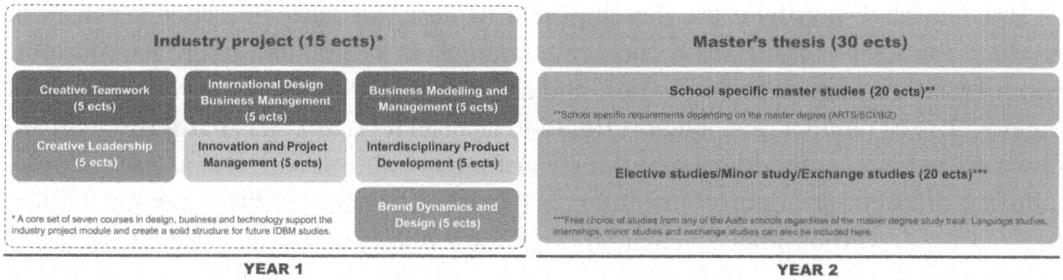

Fig. 17.2 The current IDBM master's degree structure

way'" of student-led activities, a lucid view that has enabled extensive bottom-up activity across the institution.

17.3 Educating Future Design Innovation Producers and Leaders

As we are moving into the age of intelligent technologies, where wealth in many cases is created in the spaces between specializations, society and business practice are showing an increasing demand for professionals who can act as producers of innovation, connecting the dots of the journey between initial ideas and the market. They are not only T-shaped professionals, but have a mind-set which allows for multiple and non-hierarchic contact points at all levels of art, culture, society, technology, business, data representation and interpretation, in contrast with the deterministic, arborescent model of vertical and linear connections, ill suited for today's innovation world. They have an entrepreneurial mindset linked to a transdisciplinary one, and are able to navigate cross-, multi-, inter- and monodisciplinary environments at ease.

Enabling the learning of these design innovation producers is the task of the IDBM master's program at Aalto University. The essential element in the program is the notion of initial multidisciplinarity of the participants developing into inter- and transdisciplinarity as the program advances. The program seeks complementarities between industry projects, research and learning. The degree program was built on an industry project platform, which is the key backbone for learning; a set of courses delivered by the participating schools supports the project delivery.

Crosscutting elements such as project management, design and systemic thinking, communication and presentation skills were built into the industry project content. The complementary seven courses supported the build-up of the industry project, starting with creative teamwork. The other courses deal with creative leadership, innovation management, design in business, interdisciplinary product development, business modeling and branding and market communication. The program mainly addresses product, service- and business model development.

Projects have involved the development of new business concepts, services or products, as well as translating corporate strategy to visible and tangible solutions, commercializing novel technologies, and changing existing business to suit oncoming novelty, to name a few. The original international context has naturally evolved into a global one, embedding in it the idea of cross-cultural activity (as an example, industry projects are done with emerging market partners from Vietnam, China, India, Brazil, Peru, Mexico, Uganda, and Tanzania, in addition to industrialized countries and Nordic organizations). The systemic nature of the initiatives is embedded in the very nature of the program and the idea of business development implies a future orientation and learning in strategic foresight.

17.4 The IDBM Community

In examining the strengths of the program, the first source of excellence of the program is the student body. The admission to the program is highly competitive and the level of motivation and drive that the participants demonstrate is exceptional. The focus on the master's-level studies implies that each and every student has a strong disciplinary background and in some cases extensive work experience that contribute to the shared development of interdisciplinarity in the program. The compact size of the course allows for agility in the way in which new and interesting topics can be incorporated in the offering. Finally, attention has been directed towards the creation of an IDBM community of practice (through an IDBM Club structure), which has helped to link alumni to the program in multiple ways.

Another important part of the perceived excellence of the program is very much based on the commitment of individual academics from the three universities who continuously contribute to the course offering. A key challenge is evidently identifying faculty that have interdisciplinary backgrounds. The program is managed through a collective, headed by the program director – this allows the faculty to have varying backgrounds that complement each other. The teaching practices also involve extensive external inputs from both industry and academia, linking the program in many ways to the outside world (Fig. 17.3).

As a basic principle, the program content is annually being updated and revised to reflect the changing world. The fact that different industrial partners join the program every year has a revitalizing effect on the industry collaboration, and each year the changing topics reflect the reality "out there". The overall structure and setup permits evolution and flexibility in both the delivery and development. A part of this is due to the fact that the program is not "measured to death" every year through reviews.

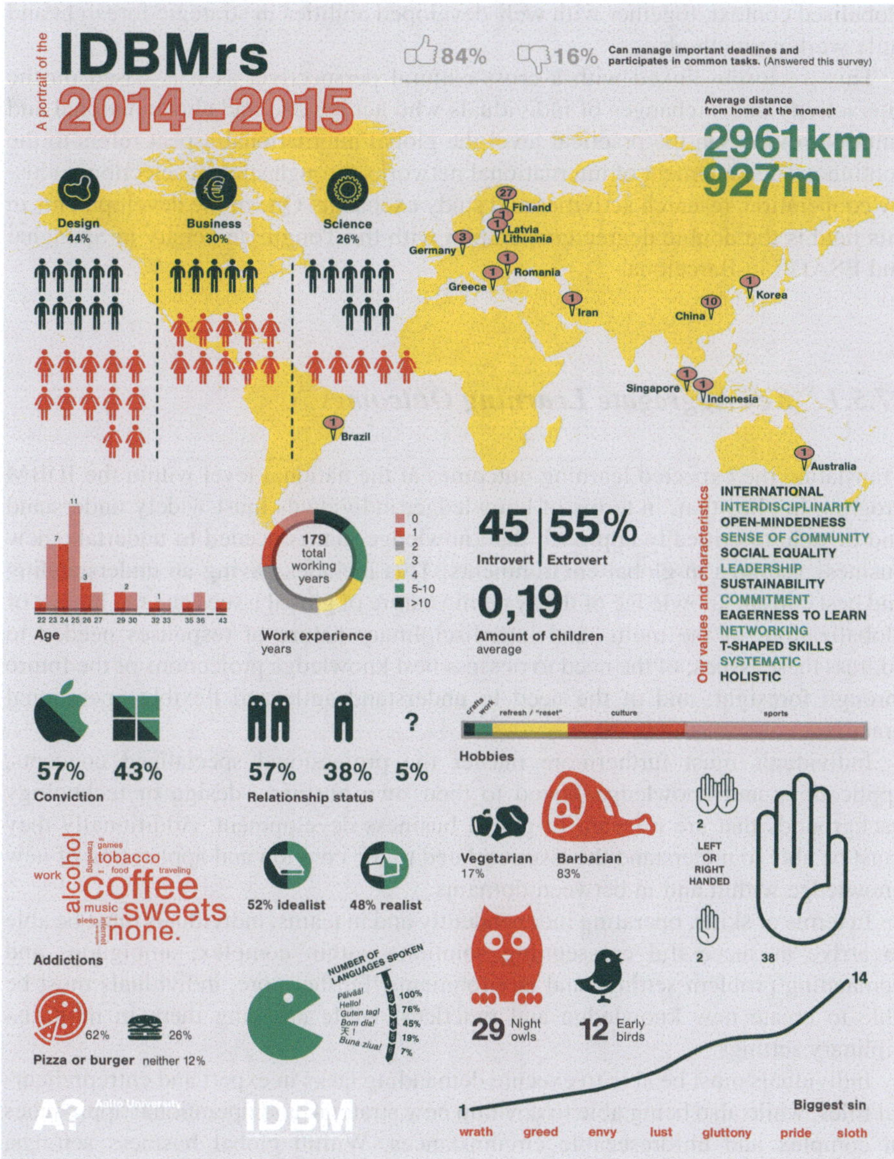

Fig. 17.3 The profile of the IDBM students 2014–15 (as made by themselves)

17.5 Towards Systemic and Integrative Competence

The overarching learning outcome of the two-year program is seen to be achievable through developing systemic and integrative competences. Achieving systemic competence is seen to require a deep understanding of interdisciplinarity in the globalised context, together with well-developed abilities in strategic foresight and agile working methods.

This is closely linked with a cross-cultural perspective, as it is based on the interactivity and exchanges of individuals who act beyond and above national and cultural groups. On the practical level the global/international aspect refers to the continuous development of international networks through student selection, industry cooperation, research activities and study exchange. One of the developments in this field is the double degree cooperation with the Tongji University in Shanghai and ESADE in Barcelona.

17.5.1 Key Aggregate Learning Outcomes

Translating the expected learning outcomes at the national level within the IDBM program implies that, in terms of knowledge, individuals must widely understand and be able to critically approach the knowledge that is needed to undertake new business ventures in global environments. This includes having an understanding and best current knowledge of the systemic nature of global issues and the impact of globalization; of the multi- and interdisciplinary nature of responses needed to address these issues; of the need to possess best knowledge projections of the future through foresight; and of the need to understand agile and flexible operational practices.

Individuals must furthermore master the professional specialized concepts, applications and knowledge linked to their own business, design or technology backgrounds that are relevant to global business development. Additionally they must be able to understand the issues related to the creation and application of new knowledge within and in between domains.

In terms of skills, operating independently and in teams, individuals must be able to arrive at successful crosscutting solutions within complex, ambiguous and demanding problem settings and environments. Furthermore, individuals must be able to create new knowledge and practices, while applying them in multidisciplinary settings.

Individuals must be able to execute demanding tasks in expert and entrepreneurial roles, while also being able to develop new strategic and operational approaches in complex and unforeseeable circumstances. Within global business settings, individuals must be able to manage and lead initiatives and other individuals; while demonstrating abilities to accumulate personal expertise and knowledge, the individual must also possess the ability to evaluate the action of self and of teams/groups and be responsible for the professional development of others. The individual must also master excellence in written and oral communication skills within and external to the professional sphere.

17.5.2 Coherence and Relevance in Learning

In interdisciplinary learning, research and industry collaboration, the issue of coherence in delivery is a key consideration. How do we ensure that the key ideas of IDBM are translated into the activities in a comprehensive and clear fashion? In this, the rhizome-enhanced T-shaped professional is the initial starting point. It is clearly recognized that the aim is not to create a designer out of a business person, or to develop a marketer from an engineer (although this may and has happened), but to create an understanding of and an ability to reconfigure to suit the tools, practices and mental models of other disciplines.

Program coherence is also related to the match between the industry projects, research and learning. Furthermore, in order to achieve coherence across the platform in systemic competence development, IDBM uses an approach consisting of five major dimensions: Tools, Environment, Management, Process, Organization (TEMPO). These dimensions are at the background of the core curriculum development, together with the professional tracks and the process Situation-Substance-Space layer, and can be used to assess external offering, with project management as a core competence that is seen to be both a means and an end. Within the industry projects, students undertake their own learning project(s).

To address the relevance of the program to students, professional orientation tracks have been identified in research, management, consulting, and entrepreneurship. It should be noted that IDBM is not (only) a research program, even though research is a vital part of the learning delivery. Nor is it an entrepreneurship program, even though it does focus on creating an entrepreneurial mindset needed in setting up new ventures within organizations or as start-ups (Fig. 17.4).

17.5.3 Industry Collaboration Projects

Running industry projects successfully over time requires that one think very carefully and consistently about the value that is created for the partnering organization. The value added for the university is clear: open-ended unique, inimitable and challenging learning environments.

The firms gain access to the consumers of tomorrow, to teams that look at the issues from various sides, with extensive recruitment possibilities. The teams can extend the range of vision of time-constrained executives and managers and they can create alternative solutions that triangulate research and development.

The industry projects are run over a period of two semesters, and the key objective is to develop product, service, and business model concepts together with the clients, aiming for meaningful solutions and innovations. The projects build on design thinking and co-creation principles. The topics are very varied, and a portfolio of projects is created for each year. In the case samples of exhibition posters of the projects, approaches were developed for promoting service design for

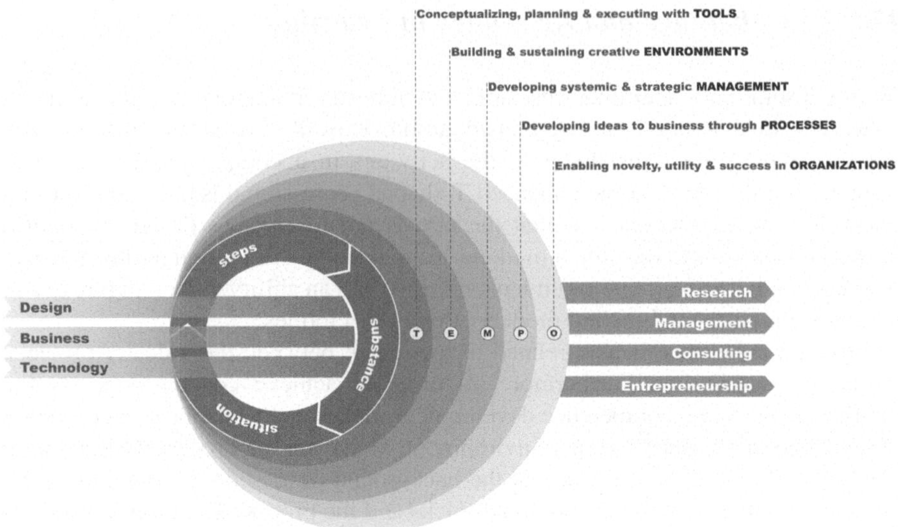

Fig. 17.4 The IDBM TEMPO matrix

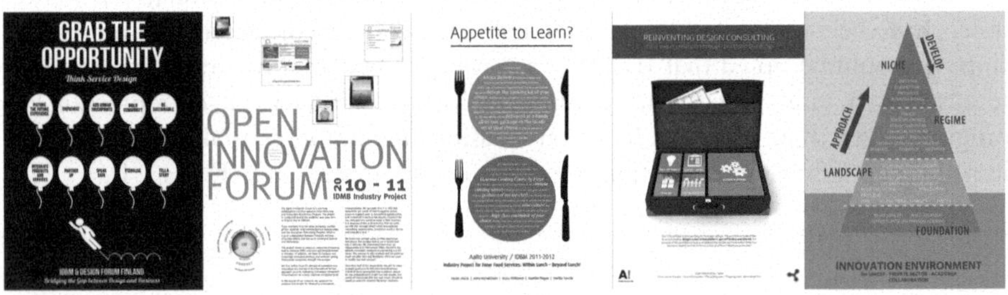

Fig. 17.5 Examples of past industrial collaboration projects

business with Design Forum Finland, a public design promotion agency (See Fig. 17.5). The Ministry of Science and Technology of Vietnam, through their the Innovation Partnership Program, wanted to develop and disseminate the concept of creating open and shared innovation platforms, while the project for Fazer Foods examined how lunch is evolving and what implications this has for business development. Pentagon Design, one of the top product and service design consultancies in the Nordic countries, wanted to rethink their business, and Unicef Uganda, together with Finland and the New York-based Unicef innovation actors inside the organization, commissioned a study on the innovation environment and its impact on product localization in Uganda, These are open source examples of projects, and the rest of the 180+ projects done to date have examined a wide range of business, technology and design topics, with many of them under non-disclosure agreements.

17.6 The Key Challenge: Scaling up?

There are over eight hundred alumni of the IDBM program currently out there, many of whom have reached significant positions in industry, some in their own businesses and others working for large and small corporations. An alumni survey in 2005 indicated that well over two-thirds of the respondents noted that the program had had a career changing effect for them, and many had found their work cut out in design-intensive businesses. Anecdotal evidence also supports the view that the IDBM alumni are often willing to hire candidates who have gone through the program, and in fact clustering around design-intensive firms seems to be happening.

A more recent study done with business alumni from 1995–2010 indicated that the respondents felt tangible financial value was created for the business organizations by the alumni. They also noted a high degree of desirability of their skills in the job market, with only 1 % unemployed after graduation. The profiles of the students were also much more international and global, and there was a more equitable gender balance in the participants. Today, almost 50 % of the students have non-Finnish backgrounds.

Evidence seems to suggest that the program has significantly contributed to the way in which design-intensive business is done in the context of Finland. This is evidently a phenomenon related to a small society: one idea, well implemented, can have a major impact. The key challenge for the future is scaling up. This seems to be happening in many ways at the moment in the Aalto context. Other multi- and interdisciplinary programs have sprung up, for example in the areas of creative sustainability, informatics, and business strategy that cut across two or more of the Aalto schools. There have also been failed initiatives, where Aalto and its predecessors have partnered in design and business, with programs ending up withering away. In other contexts, these failed programs also abound, and it seems that sustaining a long-term interdisciplinary initiative is extremely hard to do in environments where only tight disciplinary activity is rewarded through tenure and other means.

It is a unique achievement in the sphere of interdisciplinary programs to have IDBM run for twenty years continuously – it has required and will require continuous stewardship also in the future. But it seems that the time is ripe today more than ever before, as the world is becoming more and more complex, and the institutions of higher education are searching for their relevance in society. Interdisciplinary approaches are informed by the real world.

The manufacturer's authorised representative in the EU is Springer Nature Customer Service Centre GmbH, Europaplatz 3, 69115 Heidelberg, Germany. If you have any concerns regarding our products, please contact ProductSafety@springernature.com

Printed and bound by CPI Group (UK) Ltd, Croydon, CR0 4YY
25/09/2026
02222859-0015